CONTENTS

The behavioural biology of aggression

JOHN ARCHER

School of Psychology, Lancashire Polytechnic

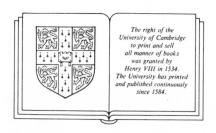

*The right of the
University of Cambridge
to print and sell
all manner of books
was granted by
Henry VIII in 1534.
The University has printed
and published continuously
since 1584.*

CAMBRIDGE UNIVERSITY PRESS

CAMBRIDGE   NEW YORK   NEW ROCHELLE
MELBOURNE   SYDNEY

Published by the Press Syndicate of the University of Cambridge
The Pitt Building, Trumpington Street, Cambridge CB2 1RP
32 East 57th Street, New York, NY 10022, USA
10 Stamford Road, Oakleigh, Melbourne 3166, Australia

First published 1988

Printed in Great Britain at the University Press, Cambridge

British Library cataloguing in publication data

Archer, John, *1944–*
The behavioural biology of aggression. –
(Cambridge studies in behavioural biology).
1. Aggressiveness (Psychology)
I. Title
302.5′4 BF575.A3

Library of Congress cataloguing in publication data

Archer, John.

The behavioural biology of aggression.

(Cambridge studies in behavioural biology)
Bibliography
Includes index.
1. Aggressive behaviour in animals. 2. Aggressiveness
(Psychology) I. Title. II. Series. [DNLM:
1. Aggression. BF 575.A3 A671b]
QL758.5.A73 1988 591.5′1 87–21792

ISBN 0 521 34558 8 hard covers
ISBN 0 521 34790 4 paperback

PN

PREFACE

Human aggression is usually viewed as a social problem, for which the behavioural scientist is expected to provide a solution. When psychologists investigated animal aggression, many of the issues chosen, and the approaches used, followed these concerns. For example, efforts were made to show that aggression can result from aversive stimulation or from frustration, and that it is subject to modification through learning. By viewing aggression in this way, some researchers came to see it as an aberrant form of behaviour. It was against this background that Konrad Lorenz's famous, but idiosyncratic, book about animal aggression, *On Aggression*, sought to re-establish the importance and significance of aggressive behaviour in the animal world.

As is now widely recognised, Lorenz unfortunately misunderstood the nature of the selective forces which have moulded aggression in the animal kingdom. He also misrepresented its motivational basis, viewing aggression as an 'appetite', a drive which built up spontaneously. Nevertheless, Lorenz's book can be understood as a valid attempt – albeit a misguided one – to reassert that aggression has a basis in the natural world, and is therefore not to be viewed simply from the perspective of a human social problem.

Like Lorenz, I shall emphasise the *biological* context of aggression in this book, but I shall do so in an entirely different way from *On Aggression*. Aggressive behaviour is viewed as a widespread solution by animals to the problems of self-preservation, protection of the young and resource competition. Examples of aggressive behaviour can be found throughout the animal kingdom, from some of the most ancient phyla, whose members possess very simple nervous systems, to those with the most complex neural mechanisms, including the human species. The specific circumstances under which aggression has evolved, and the form it takes in any one case, are the consequences of complex selective pressures acting on the animal's previous behavioural capabilities. The rules underlying these evolutionary processes are beginning to be understood, and will be discussed as one major strand throughout this book, which covers why aggression occurs in the variety of circumstances it does.

A second major strand concerns *how* aggression occurs, the mechanisms controlling behaviour. These are also complex and varied, since they serve a variety of functions in different circumstances. Mechanisms range from the simple and relatively automatic, which work well in the environment where they evolved (but poorly elsewhere), to the complex and flexible, which are sensitive to changes in the immediate environment over short periods of time, and capable of being adapted to a wide variety of conditions. Consideration of flexibility in relation to its usefulness for the individual brings us a long way from Lorenz's view, while still placing aggression firmly in the natural world.

ACKNOWLEDGEMENTS

I would like to thank Lancashire Polytechnic for granting me a year's study leave which enabled me to research, plan and write a substantial part of this book. I have been greatly encouraged by the help and advice of the original editors of the series, Pat Bateson and Paul Martin, and from Robin Pellew of Cambridge University Press, while writing the book. I would especially like to thank Paul Martin for his detailed and helpful comments on the whole manuscript, and the following who have each read and commented on several chapters: Pat Bateson, Malcolm Edmunds, Robert Elwood, Felicity Huntingford, John Rodgers and Fred Toates; also Martina Tweedie for drawing some of the figures, and Odile Archer for translating Peter Wiepkema's article.

I would also like to thank the following authors and publishers for permission to use their figures: Malte Andersson, D. Caroline Blanchard, Robin Brace, Nicholas Davies, Felicity Huntingford, Paul Leyhausen, David McFarland, John Mackintosh, John Maynard Smith, Malcolm Ostermeyer, Douglas W. Tallamy, Frederick Toates and E. C. Zeeman; American Institute of Biological Sciences, Aspen Publishers Ltd, Baillière Tindall, Blackwell Scientific Publications Ltd, E. J. Brill, Chapman & Hall, W. H. Freeman & Co., Garland Publishing Inc., Plenum Press, and John Wiley & Sons Ltd.

Special thanks to Odile Archer for her encouragement and support throughout the writing of the book.

JOHN ARCHER
January 1987

1

The history and aims of aggression research

This chapter provides an introduction to the topics covered in later chapters, by outlining the historical background to the behavioural biology of aggression, and by discussing research aims. The first part describes the origins and progress of aggression research in several different academic disciplines. The second part outlines the aims of aggression research, in relation to four different explanations of behaviour. Two of these, functional and causal explanations, dominate the research reviewed in this book, and for this reason there is a more detailed discussion of the ways in which they are related to one another.

1.1 The historical origins of aggression research

The present-day behavioural biology of aggression is woven from many historical strands. These will be considered under five headings, but the reader should bear in mind that there are many overlaps and cross-connections between them. Some of these research traditions have originated within biology and others within psychology. Even now, cross-fertilisation between the various approaches is not as great as might have been expected, or hoped for, in view of the extensive merging of ethology and comparative psychology which has occurred since the late 1950s (see Hinde, 1970, 1982).

The sub-headings that have been adopted here are as follows:
1 ethology;
2 behavioural ecology and sociobiology;
3 comparative psychology;
4 behavioural endocrinology;
5 social and developmental psychology.

Neurophysiological studies of aggression are not included in this list, although these do feature as an important input when considering issues such as the motivational basis of aggression. The neural control of aggression is a complex subject which can be viewed at a separate level of analysis from the investigation of behaviour. It was therefore decided to omit it as a major focus in this book, so as to concentrate on the already-crowded literature on the *behavioural* biology of aggression. Where neurophysiological studies are particularly relevant, they will, of course, be discussed.

1.2 Ethology

Ethology is the study of animal behaviour from a biological perspective. Typically, it involves descriptions of behaviour in a naturalistic setting, from which hypotheses about function and mechanism can be formed, leading in turn to experimental investigations.

The study of spatially related forms of aggression, notably territorial fighting, has had a long history within the tradition that gave rise to ethology. John Ray, writing in *Willughby's Ornithologia* in 1678, provided one of the earliest known descriptions of bird territories, although aggressive behaviour among birds in the spring had been known since Aristotle's time (Sparks, 1982).

The origin of the twentieth century study of territorial behaviour is Howard's *Territory in Bird Life* (1920), which formed the background for later naturalistic experiments; for example, of the compressibility of bird territories (Huxley, 1934), and of stimuli which evoke attack by a territory-holder in birds and fish (e.g. Lack, 1939; Tinbergen, 1951). Naturalistic studies of bird territories enabled both Hinde (1956) and Tinbergen (1957) to analyse the motivational basis of territorial fighting into two components, site attachment and hostility, and to speculate about the varied functions of territories.

Conder (1949) noted a second form of fighting which was also related to the opponent's spatial position: most birds will attack a conspecific which approaches too near, and the critical distance is referred to as the 'individual distance'. Marler (1956*a*) speculated that territorial behaviour may have evolved from individual distance fighting (Chapter 2).

Territorial and spatially-related forms of fighting have now been found in a wide variety of taxonomic groups (Wilson, 1975). In many cases there is a departure from the classic picture of the exclusive territory occupied by birds in the breeding season. For example, many mammals show spatiotemporal territories, in which specific localities are defended only at particular times. More recently, the study of territorial fighting has followed two lines, one involving experimental analysis in the laboratory (e.g. Mackintosh & Grant, 1966; Peeke, 1969; Mackintosh, 1970) and the other a functional approach to behaviour under naturalistic conditions (section 1.3).

At around the time of Howard's pioneering work on bird territories, Craig published two influential papers on motivation, one of them specifically devoted to aggression. In the first paper (Craig, 1918), he made the distinction between appetitive behaviour (behaviour involved in searching for a specific set of conditions) and the consummatory act (behaviour which led directly to the satisfaction of an 'appetite'); this distinction formed the basis for later conceptual schemes (e.g. Tinbergen, 1951). Craig distinguished between appetites, forms of instinctive behaviour which conformed to this pattern (e.g. eating, drinking and copulation), and 'aversions', where the animal performs an activity to remove itself from a particular set of conditions (fear and avoidance responses). In the other paper, Craig (1928) placed the motivational basis of aggression into the second category, stating that animals fight to rid themselves of the interference or thwarting of an instinct. This view, which is very reminiscent of the later frustration-aggression hypothesis (Dollard *et al.*, 1939) described in section 1.6, is one which I view as essentially correct. It contrasts with that of another eminent pioneer of the ethological approach, Konrad Lorenz.

In his well-known book *On Aggression* (Lorenz, 1966) and in his more technical paper on motivation (Lorenz, 1950), Lorenz regarded the motivational basis of aggression as essentially similar to an 'appetite' in Craig's classification. In other words, the probability of aggression was viewed as increasing spontaneously in the absence of performance of aggressive acts. Lorenz's view can be assessed indirectly through investigating the following more specific aspects: whether animals seek out fights, the influence of isolation, and whether fighting decreases the likelihood of a fight in the near future (Hinde, 1967, 1970; Johnson, 1972). The empirical evidence indicates that Lorenz's view is incorrect. In addition, if aggression *were* controlled in an appetitive way, this would make little functional sense since it would occur in response to internal events unrelated to the presence of an opponent or a provocation (Archer, 1977*a*; Toates & Archer, 1978). Despite such strong arguments against it, Lorenz's view has had a wide public impact and continues to command attention in some psychological and psychoanalytic accounts which neglect the majority of the biological evidence on aggression (e.g. Storr, 1968; Brown & Herrnstein, 1975; Siann, 1985).

The study of the function of fighting received little attention in classical ethology, with the exception of Lorenz's speculations on why animals do not engage in damaging fights (see below).

1.3 The functional approach: behavioural ecology and sociobiology

The fields of study now referred to as behavioural ecology and sociobiology have been identified as such only relatively recently, although they can be viewed historically as an amalgam of population ecology, social ethology and evolutionary biology. Brief accounts of the development of this field can be found in Hinde (1982), Trivers (1985) and Archer (1986a). Their characteristic feature is a functional approach: behaviour is studied in terms of how it contributes to the animal's survival and reproductive success (its 'fitness'). Natural selection is seen as a designing agent which changes gene frequencies according to their association with characteristics contributing to fitness.

A related field which has contributed to, but perhaps been overshadowed by, the contemporary functional approach, is that of Crook (1970b), who sought to relate interspecific differences in social structure to differences in habitat. These studies, termed 'socioecology' by Crook, provided qualitative and descriptive information, which contrasts with the mathematical models used more recently in behavioural ecology.

The pioneering work of Crook (1965, 1970a, b) and Crook & Gartlan (1966) was concerned with establishing broad relationships between social organisation and ecology, in avian and primate groups. This was followed by more detailed studies of how food requirements were related to social structure among more restricted taxonomic groups (e.g. Clutton-Brock, 1974, on two colobus species; and Jarman, 1974, on antelopes).

The basis of all contemporary approaches to the study of function in animal behaviour is the realisation that, under most conditions, selection acts on the individual rather than at a group level (Hamilton, 1964; Maynard Smith, 1964). Many speculations about the function of aggression prior to the widespread impact of individual selectionist thinking involved group-selectionist benefits. For example, Lorenz (1966) suggested that threats, bluffing and appeasement gestures in animal fights had evolved to avoid killing too many of the same species, and various authors (including Scott & Fredericson, 1951) suggested that dominance had evolved in order to prevent the widespread outbreak of fighting in animal societies.

Such speculations have now largely given way to functional hypotheses based on individual consequences. In behavioural ecology and sociobiology, these are often incorporated into formal models, contrasting with the verbal hypotheses which prevailed in earlier ethological studies of function (e.g. Tinbergen *et al.*, 1962; see Curio, 1973; Hinde, 1975).

Broadly speaking, the formal models involve either economic or game theory approaches. Economic models incorporate the assumption that animals behave so as to maximise or minimise some 'currency', such as energy intake or the time spent searching for food, which is related to their survival and reproductive success (i.e. 'fitness'). The assumptions and limitations of such models have been extensively discussed in the behavioural ecology literature (e.g. Davies, 1978*a*; Krebs, 1978; McCleery, 1978, Maynard Smith, 1978; Lewontin, 1979; Mayo, 1983).

Animal territories have been studied in these terms since the pioneering work of J. L. Brown (1964), who suggested that the occurrence of territorial defence could be predicted from the principle of 'economic defendability': it would occur where the benefits obtained from the territory exceeded the costs of defence. Brown's economic approach stimulated a number of more specific models which predict, for example, whether a bird's territorial behaviour is related either to the maximisation of daily energy intake (Hixon, Carpenter & Paton, 1983; Schoener, 1983) or to the minimisation of daily energy expenditure (Pyke, 1979).

Such economic models are generally applied to the relationship between an animal and a resource in the environment (Chapter 7). When considering social behaviour, the optimal strategy will depend on the available alternative strategies and their frequencies. These cases are more suitable for game theory models combined with the concept of ESS (evolutionarily stable strategy).

Game theory models were first used to analyse fighting strategies by Maynard Smith (1972). The first widely known paper is that of Maynard Smith & Price (1973). This approach has had a number of important consequences (Krebs & Davies, 1981): first, it shifted the emphasis from an optimal strategy in a design sense to one which is evolutionarily stable, i.e. optimal only insofar as it is better suited to current conditions than are available alternatives; secondly, the particular ESS will depend on the values assigned to the pay-offs, so that the models can predict the circumstances in which particular forms of behaviour will occur.

The paper by Maynard Smith & Price led to a range of further models of fighting strategies, as well as a number of empirical studies designed to test them (Chapter 9). This approach eventually led to a motivational model of aggression based on functionally important variables (Maynard Smith & Riechert, 1984), a fresh look at animal signals (Caryl, 1979; Krebs & Dawkins, 1984), and to some interest from comparative psychologists (D. C. Blanchard & R. J. Blanchard, 1984*b*).

The functional approach is the newest of all the behavioural approaches to animal aggression. With the possible exception of endocrinological studies, it is also the most rapidly developing field.

1.4 Comparative psychology

One subject which has featured prominently in the comparative psychology (Dewsbury, 1984*a*) of aggression is 'dominance'. This topic also crosses the boundaries of what we now call social ethology (Crook, 1970*a*), and its origin is generally attributed to the observational studies of the social interactions of chickens by the Norwegian researcher Schjelderup-Ebbe in the 1920s. This work was in fact pre-dated by the recognition of dominance in the social insects by Huber and Hoffer in the nineteenth century (Wilson, 1975).

By counting the pecks each bird gave to the others in the group and noting which one pecked which others, Schjelderup-Ebbe was able to construct a linear sequence from the one which pecked most, the dominant or 'despot', to the one which only received pecks, the subordinate (Sparks, 1982). This study provided the origin of the contemporary concept of 'dominance' which in later research came to refer not only to the outcome of aggressive interactions but, more importantly, to the priority of access to resources such dominance was supposed to confer.

It is important to realise that 'dominance' is a description of the regularities of winning or losing fights, and it is an empirical question whether or not this confers priority of access to resources. It is also an empirical question whether dominance measured by one method (e.g. by observing fights) is consistent with that measured in other circumstances (e.g. by observing priority of access to food or to a mate). Early empirical studies, such as that of Uhrich (1938) on the housemouse, soon found that dominance orders assessed in these different ways were not necessarily related. The issue of the relationship between dominance measures continues to provide controversy (Richards, 1974; Syme, 1974; Bernstein, 1981).

Much of the aggression research of the 1930s and 1940s in the comparative psychological tradition was directed towards establishing factors responsible for 'dominance' (as assessed by success in fights) in laboratory rodents (see Allee, 1942). For example, Uhrich (1938) described experiments on the influence of castration, food restriction, blinding, and colouring the opponent, on dominance hierarchies in mice. Ginsburg & Allee (1942) studied the influence of previous experience of victories on dominance in mice. This led to a series of studies on dominance and fighting in mice by Scott (1944, 1946), which culminated in a review of this area by Scott & Fredericson (1951): they summarised the influence of variables such as cold, food deprivation and castration on fighting in mice and rats. Dominance was (rightly) seen as established through learning, and the effect of previous experience on the propensity to attack was emphasised. Scott & Fredericson also noted the importance of the 'home cage' or prior residence effect: an animal will tend to initiate attack more readily and to win more frequently when it is fighting on familiar ground. This foreshadowed the following later developments:

1 the establishment of the prior residence effect in both fish and rodents (e.g. DeBoer & Heuts, 1973; Jones & Nowell, 1973);
2 linking it with naturally occurring territories in a wide variety of animals; and
3 the game theory models predicting the evolution of respect for ownership suggested by Maynard Smith (1976) and others.

At the time, however, Scott & Fredericson (1951) did not link their prior residence effect with territoriality. Indeed, Scott (1946) had earlier concluded that territoriality was absent in mice, a view later shown to be erroneous by Crowcroft (1955), Mackintosh (1970) and others.

One problem with the early attempts to study dominance in laboratory mice was the difficulty in applying a concept derived from species (such as the domestic fowl) which show a linear hierarchy to those which do not, and in which one individual (the territory holder) dominates all the others. Further problems in transferring the concept of dominance from cage to natural conditions were realised by primate field researchers such as Gartlan (1968) and Rowell (1974).

Another problem with research which viewed dominance as an important force in animal societies was the group selectionist thinking behind the term. Following Allee (1942), Scott & Fredericson (1951) remarked that 'fighting may be of use to the society as a whole, even though it may appear to be merely destructive for an individual' (p. 300). The notion that dominance provided a regulating force that prevented destructive aggression breaking out in animal societies was a prevalent one. The focus on individual competition and more precise thinking about selective advantage that has developed since the 1960s (see previous section) revealed the fallacy of such an idea. (It is interesting that Allee, in 1942, suggested a number of individual selective advantages for dominance, which have a contemporary ring about them, but then went on to suggest that group selection may be a more important force.)

Besides an emphasis on studying dominance, other features of the early comparative psychological tradition included investigation of both the inheritance of 'aggressiveness' as a trait, and the modification of fighting according to learning theory principles. Both subjects have been followed up in more contemporary research.

The inheritence of aggression has been studied, for example, by Lagerspetz (1964), Lagerspetz & Wuorinen (1965) and Ebert (1983). The importance of both classical and operant conditioning processes have been emphasised in social psychological accounts of the learning of human aggression, for example by Berkowitz (1970) and Buss (1971). The laboratory investigation of classical conditioning in animals is restricted to a few studies begun by Vernon & Ulrich (1966). Reynolds, Catania & Skinner (1963) first showed that conventional reinforcers such as food could be used in the operant conditioning of aggression. This was followed by a few other studies using conventional reinforcers, but most work in this area concentrated on the question of whether fighting itself could act as a reinforcer, an issue carried over from the earlier work, and one which is closely connected to the motivational issue of whether aggression is to be viewed as an appetite or an aversion.

1.5 Behavioural endocrinology

Behavioural endocrinology is a biological tradition which has close links with both comparative psychology and ethology. However, unlike these two purely behavioural fields, it has depended upon technological advances for its progress on the endocrinological side. Beach (1976) has remarked that progress in the analysis of behaviour has not kept up with the physiological techniques that have become available, at least in the study of sexual behaviour.

It has been known for centuries that castration of domestic animals results in docility. The first experimental demonstration that castration reduced aggressive behaviour in roosters was carried out in 1849 (Harding, 1983). The earliest systematic attempts to study how the aggressive behaviour of selected species of birds and mammals was influenced by male hormone manipulation in the laboratory were those of Uhrich, Allee and Kuo. Uhrich (1938) found that castration before the age of puberty would abolish fighting in male mice, although fighting would continue for a considerable length of time in males castrated post-pubertally. Allee, Collias & Lutherman (1939) studied small flocks of domestic fowl, and administered testosterone propionate for 7–8 weeks to birds low in the peck order. They found that the larger the dose, the more fights that would be initiated against those higher in the order. The experiments of Kuo (1960*b*), carried out in the 1930s, found a relationship between testis weight and aggressiveness in quail.

Studies of this type, on the influence of testosterone on aggressive behaviour in both laboratory rodents and domestic fowl, were continued after World War II (e.g. Beeman, 1947*a*, *b*; Allee *et al.*, 1955). Technical advances in hormonal assays during the 1960s and 1970s led to an extension of these studies to investigate such topics as the following:

1 the interaction of male gonadal hormones and experience (e.g. Lumia, Rieder & Reynierse, 1973);
2 male hormonal influences under field conditions (e.g. Watson, 1970; Lincoln, Guinness & Short, 1972);
3 primate studies of gonadal hormones and aggression (Bernstein, Gordon & Rose, 1983); and
4 studies involving a wider variety of reproductive hormones, notably the investigation of female hormones (Floody, 1983).

The majority of these more recent studies have concentrated on mammals, and this is where the most rapid advances in knowledge have been made. Nevertheless, more has become known about gonadal hormonal influences on aggression in birds (Harding, 1983), an area which at one time relied on a few isolated pioneering studies (e.g. in starlings: Davis, 1957; Matthewson, 1961; in weaver birds: Crook & Butterfield, 1968).

Knowledge of the behavioural endocrinology of other vertebrate groups has always lagged behind that of birds and mammals. For reptiles and amphibians, there is now sufficient evidence available on both the behaviour of some members of these groups and their basic endocrinology for studies of hormonal influences on aggression to be undertaken profitably (Greenberg & Crews, 1983).

Teleosts are a more widely studied group in behavioural studies, but, again, lack of knowledge of the endocrinology – and the greater technical difficulties involved in hormonal manipulations – have held the area back (Villars, 1983).

Much less is known about the hormone–behaviour interactions in invertebrates, whose hormones are substantially different from those found in vertebrates, and whose small size presents technical problems in assaying blood levels of hormones (Breed & Bell, 1983). Again, as techniques become more refined, extensive opportunities are opened up for research (e.g. see Roseler *et al.*, 1984; Roseler, Roseler & Strambi, 1986).

So far, we have only considered studies of reproductive hormones, but there are isolated glimpses of knowledge regarding other hormones. For example, thyroxine may be involved in the control of aggression in salmonids (Villars, 1983) and there is a possible involvement of the adrenals in the status-related colour changes of lizards (Greenberg & Crews, 1983). Perhaps the only systematic area of aggression research involving non-reproductive hormones has been carried out on the pituitary–adrenocortical system of mammals. Since the studies of Brain, Nowell & Wouters (1971) and Harding & Leshner (1972) in the early 1970s, the hormonal influences on, and consequences of, aggressive behaviour have been extensively investigated, notably by Brain (1977, 1979, 1981*a*), Leshner (1983*a*, *b*) and their colleagues.

1.6 Social and developmental psychology

The psychological traditions that have exerted most influence on aggression research in these areas are psychoanalysis and behaviourism. One consequence of the behavourist approach was the instigation of animal studies on issues viewed as being of central importance to the study of human aggression, notably frustration, aversive stimulation, and the learning of aggression.

The present use of the term 'aggression' in behavioural studies, meaning a 'hostile or destructive tendency or behaviour', is derived from the psychoanalytic writings of Freud and Adler, where aggression was viewed as a drive and 'aggressive' is the corresponding adjective (OED Supplement, 1972).

A group of psychologists at Yale University in the late 1930s sought to combine the extensive subject matter of psychoanalysis with the rigour and precision of the behaviourist tradition (e.g. N. Miller, 1979; P. H. Miller, 1983). In 1939, their group published the famous book *Frustration and Aggression*, which outlined the hypothesis that aggression resulted from frustration (Dollard *et al.*, 1939). This was part of a wider endeavour to operationalise psychoanalytic concepts by reformulating them in learning theory terms. In the book, they took one

Freudian notion, that of aggression originating from the blocking of a pleasurable drive, and recast it as follows: any interference with a rewarding (pleasure-inducing) activity would produce a state termed frustration, which would then lead to anger and aggression.

It soon became clear to some of the Yale group that there were problems with this rather rigid formulation in that other reactions are also evoked by frustration (Miller, 1941). Another problem with the hypothesis was one of definition. The term frustration is rather difficult to define, and there was a temptation to resort to a circular definition; that is, to find the frustrating event in retrospect when an act of aggression had occurred. Nevertheless, the frustration view has had a long-lasting influence on aggression research in psychology. For example, Berkowitz (1962) used a modification of the frustration–aggression hypothesis in his influential book on the social psychology of aggression.

One attempt to overcome the definitional problems involved with frustration was to study frustration in animals. In the studies initiated by Azrin, Hake & Hutchinson (1965a), frustration was operationalised as ceasing to reward an animal which had been on a continuous reinforcement schedule. The Finnish researchers Lagerspetz & Nurmi (1964) used a physical barrier to prevent an animal completing a previously reinforced response.

A rather different way of trying to provide an unambiguous 'animal model' of aggression was initiated by the same group of North American researchers (Azrin, Hutchinson and Ulrich). It followed an observation made by O'Kelly & Steckle (1939) that repeated electric shocks applied to the feet of male rats initiated long-lasting fighting between them. Ulrich & Azrin (1962) began a series of ethically questionable studies (see Drewett & Kani, 1981; Archer, 1986b), in which repeated shock was used to initiate fighting in pairs of rats. Owing to the Skinnerian tradition in which they were operating, many of these studies appear in retrospect to have addressed questions of little or no theoretical significance.

In the 1970s several more ethologically oriented researchers, notably R. J. & D. C. Blanchard (1977, 1981), argued persuasively that pain-induced fighting was defensive in form and motivation, and hence did not provide a useful 'model' of aggression. This controversy involves the questions of definition and classification of aggression which will be discussed in Chapter 2.

The influence of the Yale group on contemporary psychological approaches to aggression goes far beyond the frustration–aggression hypothesis, influential as this was. In their book *Social Learning and*

Imitation, Miller & Dollard (1941) began the liberalisation of S–R (stimulus–response) theory that culminated in the post-war social learning theory approach to social development. Imitation was the key learning process which was introduced to make credible accounts of social development based on learning theory principles. This process was investigated extensively by post-war researchers, notably (in relation to aggression) by Bandura (1973*a, b*). His earlier studies included the famous 'Bobo' doll experiment (Bandura, Ross & Ross, 1961), in which 3- to 6-year-old children were shown an adult carrying out aggressive acts to the doll, or an adult interacting non-aggressively, or no adult model. Children from the first group showed more aggressive acts in their subsequent play.

In contrast to the other processes involved in human aggression (e.g. frustration, aversive stimulation, classical and operant conditioning), there are no comparable animal studies of the imitation of aggression. Why animal researchers should have chosen to investigate frustration in animals and yet to ignore imitation is not entirely clear, except that a tradition of investigating frustration was already well-established (e.g. Amsel & Roussel, 1952) whereas the study of imitation was not a central concern in animal learning: indeed, its progress may have been hampered by adherence to traditional animal learning methodologies (Davis, 1973). The importance of imitation for the learning of aggression among social animals is clearly one area which requires investigation in the future.

1.7 Four questions in the study of behaviour

So far, we have seen that the present-day study of the behavioural biology of aggression has arisen from a number of traditions which have focused on some rather different issues. The two major questions being asked in behavioural studies concern function and motivation (or causation). Behavioural ecologists and some ethologists are primarily interested in functional questions. Motivation has been studied in the ethological tradition of Craig, Lorenz and Tinbergen on the one hand, and the psychological traditions of Dollard *et al.*, Scott & Fredericson, and Hutchinson & Azrin, on the other. Motivational questions are closely connected with the study of hormonal, genetic and learning influences.

Tinbergen (1963) outlined four different questions which are asked in behaviour studies. Besides the functional and causal questions we have considered so far, he also described developmental and evolutionary–historical explanations of behaviour.

There is no strong tradition of studying developmental influences in research on animal aggression. The information available is, as Bekoff (1981) recognised, isolated pieces with little or no framework to hold them together. The comparative psychological tradition included some studies of the impact of isolation-rearing on the social behaviour of dogs and rhesus monkeys, and these often included an assessment of aggressive behaviour (Mason, 1960, 1963; Harlow, Dodsworth & Harlow, 1965; Fuller, 1967; Kuo, 1967). A more specific attempt to investigate how particular periods of isolation-rearing, at different ages, would affect adult aggression in rodents was begun by King & Gurney (1957) and King (1957), and has led to spasmodic studies up to the present time (e.g. Cairns, Hood & Midlam, 1985).

Tinbergen's fourth question concerned evolutionary antecedents, tracing the pathway from one form of behaviour to another through evolutionary history. This can be carried out only indirectly, principally by studying closely related species. The classic studies are those of Lorenz (1958) on the evolution of duck displays, but the same general approach has been applied to many other forms of behaviour, including human smiling and laughing (van Hoof, 1972). Aggressive displays can also be studied in this way, but the historical question has generally been relegated to incidental discussions in the literature on aggression (with some exceptions: e.g. Geist, 1966, who studied horn-like organs in ungulates in relation to the evolutionary history of their use in fighting).

The main questions which have been addressed in studies of animal aggression have therefore concerned its evolutionary function(s), and its motivational or causal basis.

1.8 Functional and causal questions

At present, functional and causal approaches to the study of behaviour represent quite different questions, yet they have often been confused. Since there has been little or no systematic attempt to integrate the two questions, this issue will be discussed here to provide a framework for later chapters where the interrelationship between function and motivation will be a recurring theme.

Earlier I referred to the biological definition of function as those consequences of behaviour which enhance fitness. A second use of the term is a wider one, to denote the task or job a system is 'designed' to perform. This meaning was familiar to biologists long before the time of Darwin (Mayr, 1982). In contemporary behavioural ecology, specific optimality models begin with the general assumption that fitness is being maximised and then seek to show in more specific design terms how this is achieved.

We turn now to the question of whether biological function, the adaptive consequence of behaviour, is linked with causation, the immediate control of behaviour. Logically, the two issues are unrelated. As long as a particular mechanism *works*, i.e. produces the required functional end-point, it does not (in principle) matter *how* it works. There are many examples illustrating how this distinction is maintained in practice. Classical ethological studies have emphasised relatively stereotyped responses to specific stimuli, which occur whether or not the response makes functional sense in that particular context. A male robin will attack a bunch of red feathers on a stick (Lack, 1939), even though such a response has no beneficial consequences.

Similarly, mammalian male sexual behaviour involves a controlling mechanism which operates whether or not its functional end-point (fertilisation) is likely to be achieved. The mechanism seems to be based on positive feedback resulting from a particular type of stimulation. Ejaculation occurs after a build-up of stimulation to a level unlikely to be achieved outside copulation. The mechanism 'works' (in a functional sense) on the basis of this probabilistic coincidence (Toates & Archer, 1978). In behavioural ecology, the term 'rule of thumb' is used to describe the way in which a mechanism achieves a result good enough to fulfil its biological function, yet operates on principles logically unrelated to this function.

In contrast to these examples, the control systems for many forms of behaviour may contain representations of a desired end-point. This not only introduces more flexibility into the control of behaviour, but also allows expected consequences of behaviour (related to function) to become part of the controlling mechanism. Powers (1978) argued that this has occurred repeatedly in the course of behavioural evolution. The same principle has been used, in discrepancy models, to explain a range of different types of behavioural control. Here, a large difference between the input and an internal representation produces action which subsequently reduces the discrepancy. The internal representation generally corresponds to a beneficial consequence. Such models have been applied to imprinting (Salzen, 1962), infant distress (Kagan, 1974), displacement activities (McFarland, 1966), human emotions (Duval & Wicklund, 1972), as well as to aggression (Archer, 1976a; Wiepkema, 1977). In other cases, such as the development of mate choice in quail, the internal representation does not itself correspond to a beneficial consequence but bears a constant relationship to it (Bateson, 1978, 1980, 1982).

In many cases, neither a stereotyped response mechanism, nor one which incorporates a general representation of the desired end-point, will ensure that behaviour leads to beneficial consequences. Where the likelihood that behaviour will be adaptive depends on variations in the external conditions, decision-making mechanisms which enable the animal to assess the likely consequences of behaviour will be required. A number of different types of behaviour have now been shown to be controlled by decisions based on an assessment of the costs and benefits of that behaviour. Food foraging under natural conditions (Krebs, 1978; Krebs & McCleery, 1984), and operant schedule-controlled behaviour under laboratory conditions (Lea, 1978; Staddon, 1984) were the first to be studied in these terms. McFarland (1976, 1977) and Staddon (1984) have suggested that animals' decision-making – i.e. what determines which of several alternative types of behaviour has priority for the animal at a given moment – is determined by such a cost–benefit assessment.

In the field of aggression research, the territorial behaviour of many birds has been shown to follow an assessment of environmental cues (e.g. Gill & Wolf, 1975; Carpenter & Macmillan, 1976), and so has the occurrence or non-occurrence of parental aggression in some anurans (Wells, 1981). Assessments have also been shown to affect decisions about how to proceed against a potential opponent in spiders (Riechert, 1979; Austad, 1983), hermit crabs (Dowds & Elwood, 1983), and other species.

Control system and decision-making mechanisms may be combined to form flexible control of behaviour. Assessment of relevant cues in the environment may modify the representation of the desired end-point over a short time period, and hence introduce flexibility which would not have occurred with a more stable end-point.

Powers (1973) suggested that behavioural control systems are arranged hierarchically with the output of the higher-order system providing the reference value for the lower-order system. In the present context, either control may be exerted by the higher-order system (e.g. the animal behaves so as to achieve a longer-term goal such as maintaining attachment to a mate or a territory or nest-site), or it may be uncoupled from such a high-order system, so that the animal behaves so as to achieve a series of short-term goals: these would be determined by assessment of the prevailing conditions (such as obtaining the maximum number of copulations or rate of food intake), irrespective of such long-term goals as attachment to a mate or to a nesting site. Davies & Houston (1984) gives examples of these strategies in relation to the territorial behaviour of birds.

The distinction between higher-order and lower-order control of behaviour corresponds to a similar one made by Lea (1978) to describe the relationship between operant behaviour and the costs of obtaining reinforcement. In some cases, responses are maintained despite increases in the costs of obtaining the reward: Lea applied the economic concept of an elastic demand curve to these instances. In other cases, responses are soon abandoned when costs are increased, and this was referred to as an inelastic demand curve. The two cases correspond, respectively, to higher-order and lower-order control in Powers' hierarchy.

In conclusion, some causal mechanisms will be relatively insensitive to changes in the adaptive consequences of behaviour, whereas others will be designed in such a way as to be more responsive to change. Waddington (1957) distinguished between adaptive changes which occur over three different time-scales, short-term (behaviourally and physiologically mediated), medium-term (life history) and long-term (phylogenetic). Animals which relied only on phylogenetic change would be limited in the extent they could avoid or respond to environmental change, yet environmental change is subject to no comparable limits. Supplementary adaptive mechanisms – operating within the lifetime of an individual (developmental flexibility) or over shorter time periods (e.g. learning) – enable each individual to adapt to its specific circumstances (Plotkin & Odling-Smee, 1979). This distinction between the time-scales of adaptive flexibility will be used as a framework for discussing mechanisms controlling the diversity of animal aggression in later chapters of the book. Where possible, phylogenetic adaptations such as hormonal influences on aggression will be discussed separately from the influence of the rearing environment and from mechanisms producing shorter-term changes.

1.9 Future research directions

In this chapter, I have introduced the various traditions associated with research on animal aggression, and discussed the main questions which are asked within these traditions. Although some specific areas of ignorance were pointed out, perhaps a more important task for the future is the *integration* of different approaches and aims. For example, aggression research in comparative psychology and in behavioural endocrinology would benefit from a recognition and appraisal of the rapid developments in the fields of behavioural ecology and sociobiology (Archer, 1985a). The study of aggression in the laboratory, using techniques such as pain or frustration, would benefit from setting such techniques into a more naturalistic framework. In more general terms, a consideration of function may assist the study of mechanisms, and an appreciation of mechanisms can enrich the study of function.

Considering specific omissions in the present research, it is inevitable that certain research topics will attract a great deal of interest while others will be relatively neglected. This will occur, in part, as a result of technical restrictions (e.g. as in the case of the behavioural endocrinology of insects), but a more widespread reason will be tradition and fashions within the academic disciplines. The comparative psychologists' choice of the white rat was apparently a fortuitous one (Beach, 1950) but, once it had begun, it was self-perpetuating. In a similar, but less drastic way, minor research traditions have been built up around specific techniques for producing aggressive behaviour (such as frustration, shock, or territorial introduction), involving particular animals (easily obtainable species such as mice, rats, domestic fowl and Siamese fighting fish).

Inevitably, such an approach to research will have led to some gaps which are obvious in retrospect; important topics have been neglected because they did not fit the prevailing approaches. Earlier, we noted that the role of imitation in the development of aggressive behaviour appeared to be such a topic. In a broader sense, so is the whole issue of the development of aggression. Apart from isolation-rearing studies, which were designed to demonstrate the importance of normal social experiences for the emergence of social behaviour, developmental influences have been little studied. In particular, this contrasts with the greater attention paid to more immediate influences on behaviour such as hormones, reinforcement history and external circumstances.

These specific issues will be discussed in later chapters. In the next chapter, aggression is considered in a general evolutionary framework.

Further reading

Hinde, R. A. (1982). *Ethology*. Oxford: Oxford University Press.
Johnson, R. N. (1972). *Aggression in Man and Animals*. Philadelphia: Saunders.
Svare, B. B., ed. (1983). *Hormones and Aggressive Behavior*. New York: Plenum.

2

A functional framework for the evolution of aggression

2.1 'Aggression' in sessile invertebrates

In this chapter, we consider aggression in a broad evolutionary context. Viewed in this way, it is the functions of the behaviour which assume importance, the precise mechanisms varying considerably in different taxonomic groups.

'Aggression' as a functional category is perhaps best viewed as a range of behavioural solutions to certain types of problems animals encounter in their relationships with their enviroments. The same types of problems – principally how to compete actively for resources and to fend off danger – can be seen as recurring again and again in different animal groups. Each time a particular group evolved aggressive responses to deal with such problems, this would have entailed modifying their previous neural and effector systems, so that they become capable of inflicting damage on another animal. In some cases, this would have begun by using existing systems adapted for other purposes – what Gould & Vrba (1982) termed exaptations. For example, the coral *Scolymia lacera* kills the tissues of a closely related species *S. cubensis* by using an extracoelenteric feeding response; this resembles the ectodermal feeding method it uses to digest food particles which are too large to be transported into the stomodaeum (Lang, 1971). It seems that such 'aggressive' responses evolved – mainly in the suborder Faviina – to counteract other forms of competitive mechanisms found among corals, in particular the rapid growth and expansion of ramose and foliose corals (Lang, 1973).

In other cases, the animal harnesses structures already used for locomotion in a new way. For example, some intertidal molluscs such as limpets and chitons show spatially related aggression; this is achieved in the case of the chiton *Acanthopleura gemmata* by crawling over the rival

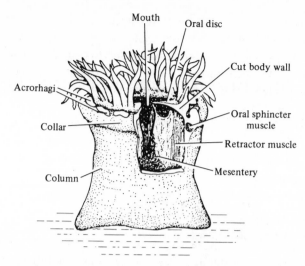

Fig. 2.1. Diagram to show the gross morphology of *Actinia equina*. Part of the column and oral disc have been cut away to display some features of the internal anatomy. (From Brace *et al.* (1979), p. 554, with original caption.)

conspecific with its foot and trying to dislodge the rival from its crevice in the rock by backward and forward movements (Chelazzi *et al.*, 1983).

Structures specialised for intraspecific aggression are, of course, found throughout the animal kingdom. Examples can again be found in sessile invertebrates which would not normally be associated with the label 'aggressive'. Brace & Pavey (1978) and Brace, Pavey & Quicke (1979) investigated aggressive responses in a solitary anemone, *Actininia equina*, which possess structures bearing batteries of nematocysts (the acrorhagi), used solely for offence (Fig. 2.1). Anemones, which were firmly attached to a substrate, were moved towards one another until their tentacles touched; this tactile contact resulted in an aggressive response in the majority of cases. (Francis, 1973, reported a similar finding for *Anthopleura elegantissima*.) Clear winners and losers could be distinguished in these disputes, in that the loser became wholly or partly closed or could become detached from the substrate. Retaliation sometimes occurred, but the initial aggressor usually won the encounter. The larger anemone was more likely to be the initiator and also to win. As we shall see in Chapter 9, this relationship has been found throughout the animal kingdom.

The mechanisms by which anthozoan coelenterates such as sea anemones and corals detect the presence of conspecifics is probably chemical and/or tactile, and the neural control of the behaviour would be simple in animals whose nervous systems consist of nerve nets. In Brace & Pavey's study, the response was found to be reflexive in nature, in that once it had been initiated, it carried on in a stereotyped form until it was completed (i.e. there is a coincidental relationship between function and mechanism in this case: see Chapter 1).

These examples show that behaviour which can be regarded as 'aggressive' in functional terms occurs even in those invertebrates which operate on a simple level of neural organisation, possess little or no powers of locomotion, and lack specialised effector organs.

2.2 A functional classification of aggression

If 'aggression' is, as has been suggested, best viewed as a range of solutions to certain problems, can we provide a useful functional classification of aggression in terms of the particular problem addressed by that type of aggression? In this section, it is argued that there are two broad functional strands in the evolution of aggression, one concerned with resource competition and the other with reactions to danger. Although these functions are distinct, the mechanisms which underlie them have come to overlap.

Most functional discussions of aggression, other than group-selectionist ones, view resource competition as the only important aspect (e.g. Brown, 1964; Wilson, 1975; Clutton-Brock & Harvey, 1976). Aggression is seen as a specialised form of resource competition (MacArthur, 1972; Wilson, 1975). In a general sense, successful competition enables an animal to achieve ends such as occupying space necessary for growth (e.g. the corals discussed above), eating more food and breeding in a more advantageous environment than a competitor (usually a conspecific). In plants and sessile animals, these ends will usually be achieved over relatively long time-scales, for example when a plant outgrows and chokes a competitor, or when a barnacle overgrows a competitor on an attachment site (Connell, 1961). In such cases, the unsuccessful competitor is denied access to the resource. When the same result is achieved over short time periods by behavioural competition, the behaviour involved is usually termed aggression if it involves active interference with the rival. (There are other forms of behavioural competition which occur over short time periods, notably reproductive interference such as the removal of a rival's sperm: see Chapter 7.) This definition of aggression is, therefore, partly functional (in terms of resource competition) and partly descriptive (it occurs rapidly, involves active interference with a rival, and often involves specialised neural and effector systems).

In some of the simple invertebrate examples, competition – for a space to live or a secure attachment site – is fairly direct. But in many animals, competition for resources occurs indirectly in the form of feeding or reproductive territories. Wilson (1975) has pointed out that, if no other factors are involved, it is energetically more efficient to compete a relatively few times for an area which contains food than to compete for each item of food. As we shall see in Chapter 8, many other factors are involved, so that territorial behaviour in defence of food occurs only under certain environmental conditions.

If we consider some of the other forms of aggression introduced in the previous chapter, two which are found in vertebrates have the clear function of obtaining and keeping resources. These are frustration-induced and instrumental aggression.

Frustration-induced aggression occurs in response to the absence, inaccessibility or delay of an expected reinforcing event. It will therefore occur in situations where there is direct competition for resources, for example where there are localised clumps of food (Chapter 7). This is a case where a functional end-point (obtaining a resource such as food) has been incorporated as an internal representation of what is expected: a discrepancy from this expectation will activate an aggressive response. Instrumental aggression – aggression which has previously resulted in reinforcement (Chapter 1) – would also operate in conditions of resource competition. In this case, the mechanism works on the basis of a learnt internal representation that links the action with the functional consequence, but this would also be subject to assessments of the schedule contingencies at the time (i.e. assessments of the costs and benefits of the choice).

The second broad strand in the evolution of aggression concerns its defensive function, its contribution to the removal of immediate danger. This can be regarded in general terms as similar to the evolutionary background of fear behaviour, which results in the mitigation or avoidance of danger, or escape from it. In an earlier consideration of this subject (Archer, 1979), I started with the premise that all organisms live in environments where physical damage is a possibility and that they can, to some extent, counteract this by repair and regeneration. An additional way of counteracting danger is by withdrawal into a protective covering, or removal of the animal from the influence of the potentially damaging stimulus: this is presumably of ancient evolutionary origin in the animal kingdom, since it can be observed in Protozoa, such as *Paramoecium* (Jennings, 1906). In metazoans, tissue damage can be monitored by nociceptors, and effector systems controlling protective and escape responses are linked to such receptors. The function of these systems is to enable the animal to avoid the influence of a damaging stimulus or to move away from its vicinity. However, if the harmful stimulus is small and localised, effector systems aimed at removing it would be more advantageous, as the animal would not have to leave the area. Protective aggression could serve such a function, and simple versions of it are represented by the counter-attacks seen in the sea anemone studies described above (Brace & Pavey, 1978); presumably, these occur in response to nociceptor stimulation and are the forerunners of the pain-induced aggression of vertebrates mentioned briefly in the first chapter, and introduced more fully in Chapters 3 and 4.

There would, however, be advantages for an animal which could respond to potential danger before physical contact. This would entail a more complex level of neural organisation than a response to nociceptor stimulation. In particular, neural representations of the environment would be used to predict events likely to cause damage. Thus, the principle that unaccustomed tactile stimulation readily evokes aggression (see above) would be extended to include intrusions into a spatial area around the body, i.e. the individual distance reactions shown by most birds and mammals (Conder, 1949; Hediger, 1950). Animals may also come to defend areas located in space, independently of their bodies, where they are relatively safe from physical danger. These would form the type of territory whose function is to provide roosting positions or shelters only (Wilson, 1975), or a refuge from predators (e.g. Stamps, 1983).

Another way of responding to potential danger in advance of tissue damage is for the animal to use representations of the temporal correlations in its own specific environment to respond to stimuli which predict noxious events. This, of course, is the basis of classical conditioning, and studies on rats have demonstrated that pain-induced attack can be conditioned in this way (Creer, Hitzing & Schaeffer, 1966; Vernon & Ulrich, 1966; Lyon & Ozolins, 1970).

One form of aggression which can be viewed as an extension of protective aggression is parental aggression, which refers to aggression whose function is to protect the eggs or offspring (cf. use of the term by Thresher, 1985, to refer to aggression by parents *to* their young). Parental aggression has often been referred to as 'maternal aggression' (e.g. Moyer, 1968; Archer, 1976a; Svare, 1981), since it has only been studied experimentally in mammalian species where it is shown by pregnant and lactating females. Some form of parental aggression is, however, found among a wide variety of animals which show parental care (Ridley, 1978; Eisenberg, 1981). In functional terms, protection of an animal's own offspring from predators can be seen as a fairly straightforward extension of defence of the animal itself. We should also expect it to be particularly pronounced in species where there is a risk of cannibalism or infanticide towards strange offspring (e.g. Bygott, 1972; Bertram, 1975; Hrdy, 1979; Elwood, 1980; Townsend, Stewart & Pough, 1984).

Although three functionally different types of aggression – competitive, protective and parental – have been outlined, these functions will, in some cases, overlap. This functional overlap applies particularly to spatially related forms of aggression.

Marler (1956*a*, *b*) suggested that individual distance provides the evolutionary precursor of territory. If this is correct, the evolution of territorial behaviour would have occurred when, first, it became more advantageous to defend a particular area located independently of the animal than an area immediately surrounding it, and secondly, the animal's central nervous system was able to form map-like representations (O'Keefe & Nadel, 1979). The second of these presumably determines the broad distribution of territorial responses throughout the animal kingdom, mainly in arthropods and vertebrates (Wilson, 1975). The first determines their detailed occurrence in particular species (see Chapters 7 and 8).

There are two possible ways in which individual distance may have evolved into territorial behaviour. The first would occur in those territories which provide only a protective function (see above); as well as defending the area around itself, the animal comes to defend an area, located independently of itself, where it is relatively safe from physical danger. Territoriality which serves a competitive function can also be seen as an extension of individual distance responses, since these would aid an animal's ability to keep a resource in its immediate vicinity when its space was invaded by a competitor. Where the ecological conditions are appropriate (Brown, 1964), resource defence could become extended to cover an area containing the resources, rather than only the area surrounding the animal's body.

Because the function of parental aggression is to preserve the life of an offspring rather than to acquire a resource, it can be regarded as a form of protective rather than competitive aggression. However, in species where there is a breeding territory – for example, in many fish, amphibians and birds – it will occur within the broad context of territorial behaviour which also serves competitive and protective functions. It has often been argued (e.g. Trivers, 1972; Ridley, 1978; Gross & Shine, 1981) that in animals with external fertilisation, prior defence of the territory will be linked to subsequent parental care by the sex which was the territory-holder.

Whether or not it is functionally distinct from territorial defence, parental aggression would appear to involve mechanisms incorporating representations either based on a spatial area containing the eggs or young, or on the eggs or young themselves. These representations are then used in a discrepancy-activated system to respond aggressively to sudden changes in the surroundings. (Studies of the parental aggression of rodents, reviewed in Chapter 6, broadly fit this description.)

In animals which form complex spatial representations, these appear to be used in the pursuit of a number of functional end-points, which may be protective, as in the case of roosting territories or parental aggression. Alternatively, they may involve competition for food and for reproductive resources, in addition to serving protective functions.

Most recent discussions of the functions of territorial aggression have emphasised resource competition (e.g. Davies, 1978*a*; Huntingford, 1984*a*), although the protective functions were covered in the earlier works of Hinde (1956) and Tinbergen (1957). However, the functions of resource competition and safety can, in some senses, be seen as extensions of one another. If an animal lives in a world which is potentially unsafe, the ultimate extension of responding only to imminent life-threatening danger is to seek to remove anything that will produce a lowering of fitness in the future, including a rival for a resource which is in short supply. Thus, in terms of contribution to fitness, there is no sharp dividing line between protection and competition.

2.3 The relationship between functional and causal classifications of aggression

In this chapter, the evolutionary origin of aggression has been discussed in terms of two broad but overlapping functions. Some mechanisms underlying aggression appear to be more closely related to one function than the other, whereas others are a mixture of the two. For example, frustration-induced and instrumental aggression are both competitive in function, whereas pain-induced attack is protective. Other types of aggression, particularly the heterogeneous category of spatially related

Table 2.1. *Moyer's (1968) classification of aggressive behaviour in animals (from Archer & Browne, in press)*

1 Predatory aggression: in response to a natural object of prey

2 Intermale aggression: in response to the proximity of an unfamiliar male

3 Fear-induced aggression: shown by a confined or cornered animal, and preceded by escape attempts

4 Irritable aggression: in response to a broad range of external circumstances, such as pain, frustration and deprivation, and shown towards inanimate as well as animate objects

5 Territorial defence: in response to an intruder into an area in which the individual has established itself

6 Maternal aggression: in response to the proximity of an animal which is perceived as threatening the young

7 Instrumental aggression: a learned response to obtain reinforcement

8 Sex-related aggression: in response to competition for a mate

aggression, is best viewed in terms of combination of functions (see above).

We now turn to the question of whether a particular *function* is reflected in the *form* of aggression, i.e. whether competitive aggression is most likely to be 'offensive' in its appearance and protective aggression 'defensive'. This issue has assumed some importance in relation to the classification of aggression.

When it became apparent that aggressive behaviour could differ in terms of its behavioural form, its function and the way it was influenced by immediate internal and external factors, several attempts were made to provide a classification of aggression (see Archer & Browne, in press). The earliest and perhaps most influential one was that of Moyer (1968), who listed eight different types of aggression, based largely on laboratory studies (Table 2.1).

Moyer's scheme can be criticised on the following grounds (Archer, 1976a):

1 so-called predatory aggression is so motivationally and neurally different from other forms of aggression that it is most usefully considered as a separate form of behaviour;

2 the criteria for classifying the types of aggression were inconsistent;

3 the neural evidence did not support a distinction between intermale, territorial, 'maternal' and irritable aggression.

Although these and other criticisms of Moyer's scheme have been raised (e.g. Huntingford, 1976*b*), it has remained widely cited in the absence of an acceptable alternative.

Thompson (1969) made the useful distinction between unconditioned aggression (e.g. resulting from pain or from frustration) and conditioned forms, resulting from either classical or operant conditioning. Such a distinction, although important in animal studies, assumes even greater significance in relation to human aggression (Feshbach, 1964).

The research of the Blanchards and their co-workers in the 1970s showed a clear behavioural and motivational distinction between offensive and defensive forms of attack in the rat and mouse (see Chapter 4). The former included certain distinct behavioural acts and a restriction of bites to the dorsal surface of the opponent. It would typically occur when a territory-holder attacked an intruder. Defensive attack again shows a distinctive form, including the occurrence of the upright boxing posture, and typically involves an intruding rat being attacked by a territory-holder; it also occurs in response to repeated painful stimulation, as used in laboratory investigations of shock-induced aggression (Chapter 1).

In seeking to offer a classification of aggression that would meet some of the objections to Moyer's scheme, Brain (1981*b*) outlined five categories of 'aggression'. Two of these concerned predation and infanticide, which are outside the scope of this book (and, I would argue, are best not included as forms of aggression). The other three categories correspond broadly to the functional groupings outlined above; Brain called these social conflict, parental defence and self-defensive aggression. The latter category included aggression in response to pain, individual distance intrusion, and attack by another animal when escape is blocked. Antipredator attack, which is very similar in form to pain-induced attack in the rat (R. J. Blanchard, 1984), should also be included. Brain argued that these forms of aggression occur in response to a threat, are fear-motivated, are generally preceded by avoidance attempts and their target sites are not restricted (as they often are for offensive attack). Similar conclusions about a basic division of aggression into offensive and defensive forms have been reached by other researchers using behavioural or neurophysiological criteria (e.g. R. J. & D. C. Blanchard, 1977, 1981; Adams, 1979, 1980; Rasa, 1981; Rodgers, 1981*b*). However, Brain's classification went further in implying a functional as well as a causal grouping: self-defence behaviour occurs in response to a direct threat to the animal's physical well-being.

Although I have argued that aggression can be divided into functional categories rather similar to those proposed by Brain, my functional groupings are not intended to imply that any particular motivational state or form of behaviour is associated with that group. In contrast, Brain appears to equate each functional grouping with a specific type of behaviour, either offensive or defensive behaviour, as indicated above for the rat. Instead, I would argue that aggression which is protective in a functional sense can, in theory, involve offensive or defensive aggression (or both), defined on the basis of its form, motivation and physiological basis. Which one occurs in any particular case will depend on a variety of factors, ranging from species adaptations to situational variables. For example, although rats show a characteristic defensive reaction when cornered by a predator, large ungulates such as the moose show a more offensive form of antipredator attack (Geist, 1966, 1974a; Kruuk, 1972). In cases where pain is more moderate and restricted than in shock-induced aggression studies, offensive attack can occur (Lloyd & Christian, 1967; Cairns *et al.*, 1985, for the housemouse; Poole, 1974, for the polecat; Jacobsson, Radesater & Jarvi, 1979, for a cichlid fish).

Similarly, other forms of aggression – irrespective of their function – can contain elements of both attack and defence depending on the species and circumstances. Hinde (1956) and Tinbergen (1957) recognised this point in their reviews of the territorial behaviour of birds: the tendencies both to attack *and* to flee are associated with territorial fighting, and selection favours not one or the other in isolation but their relative strengths.

2.4 Conclusions

In this chapter, we have examined the evolutionary context and functions of aggression. The distinction between 'protective', 'parental' and 'competitive' forms of aggression seems to provide a useful basis for classifying different types of aggression on functional grounds, as long as it is clear that these are functional groupings only, and each encompasses a variety of forms and mechanisms. This functional classification will be used to discuss the diversity of aggression in the following six chapters. There are two chapters on each of the three forms of aggression. The first chapter will be devoted to a functional analysis of diversity in aggression shown by different species, the two sexes and over the life history. The second chapter will consider the mechanisms underlying such diversity, first in terms of the ways in which function and mechanism may be related (Chapter 1), and secondly in terms of how phylogenetic, ontogenetic and short-term adaptive mechanisms can produce the diversity of animal aggression described in the functional chapter.

Further reading

Brace, R. C., Pavey, J., & Quicke, D. J. L. (1979). Interspecific aggression in the colour morphs of the anemone *Actinia equina*: the 'convention' governing dominance ranking. *Animal Behaviour*, **27**, 553–61.

Brain, P. F. (1981*b*). Differentiating types of attack and defence in rodents. In *Multidisciplinary Approaches to Aggression Research*, ed. P. F. Brain & D. Benton, pp. 53–78. Amsterdam: Elsevier/North Holland.

Wilson, E. O. (1975). *Sociobiology: the new synthesis*. Cambridge, Mass.: Harvard University Press.

3

Protective aggression:
a functional analysis

In the previous chapter, I introduced 'protective aggression' as a functional category which includes responses to a predator, attack by a conspecific, sudden pain, and individual distance intrusion. Protective aggression is the subject of this chapter and the next. Here, we are concerned with functional issues, notably the selective pressures which have produced variations between species, the two sexes, and at different stages of the life cycle. In Chapter 1, we briefly considered different approaches to the study of function, contrasting verbal ethological hypotheses with the more formal models used in behavioural ecology and sociobiology. There are few, if any, formal models relevant to protective aggression, but a range of verbal hypotheses have been advanced.

In the first section, we consider interspecific variations in the form and intensity of protective aggression, and in the second the relationship between protective and competitive aggression, together with the diversity of protective aggression shown to different predators. We then consider whether a functional approach can provide an explanation for sex differences in protective aggression which has been noted in laboratory studies of rodents. Finally, we show how functional considerations might be related to developmental changes in protective aggression.

3.1 Variations in the form and intensity of protective aggression

Protective aggression is widespread throughout the animal kingdom, from the sessile invertebrates described in the previous chapter to the mammals. Sea anemones, for example, show aggressive responses when attacked (Brace *et al.*, 1979). Territorial limpets show different forms of aggressive response to a conspecific intruder than to a predator

(Stimson, 1970). The praying mantis may attack a predator, by striking it with the clawed forelegs, or by biting (Edmunds, 1972).

Interspecific variation in the form of protective aggression can be understood in terms of predation pressure (cf. Adams, 1980). A similar case was argued for fear and avoidance responses (Archer, 1979): it was suggested that their form in different species is related to factors such as the frequency of encounters with predators, and whether escape is possible. In this context, it was also noted that some animals possess morphological structures capable of inflicting injury upon predators which attempt tactile contact: examples include the spines of hedgehogs, porcupines and sea urchins (Edmunds, 1974). A logical extension of the possession of such structures is for an animal to attack the predator, to drive it away either completely or sufficiently to enable escape to occur.

Attack as an antipredator strategy may take a number of forms, and can involve group or individual behaviour. It includes some forms of mobbing in birds (Harvey & Greenwood, 1978), attacks on snakes by rodents such as *Cynomys* (Owings & Loughry, 1985), and the standing and fighting shown by large-bodied species of mammals such as the moose (Geist, 1966, 1974a), the giraffe (Edmunds, 1974), the oryx (Cloudsley-Thompson, 1980), the rhinoceros (Kruuk, 1972), and the hamadryas baboon (Kummer, 1971). These examples are all drawn from cases where the animal or animals are physically able to drive off a predator as a result of their size or numbers (Huntingford, 1976b). There are also examples of small but potentially dangerous animals, such as snakes, which show counter-attack to predators (Arnold & Bennett, 1984). In addition, a number of species show bluff displays which are apparently effective in removing predators (Edmunds, 1974). For example, praying mantids show a dramatic startle display when attacked by avian predators (Edmunds, 1972, 1976). Young owls bill-snap and fluff up their feathers when threatened, but do not bite (Edmunds, 1974). The aardwolf *Proteles* shows a threat display similar to that of other carnivores, but its mouth is kept closed; evidently this is because its dentition is extremely reduced and to open its mouth would reveal the inadequacy (Ewer, 1968)!

Animals nesting with their young, and which have no other strategy available but to confront and attack a predator, will show similar forms of antipredator aggression even in cases where they appear not to be physically or numerically well-matched. (This is discussed separately in Chapters 5 and 6, as parental aggression.)

The examples mentioned so far are all 'offensive' in form and are aimed at driving off or injuring the potential predator. Baboons, for example, may kill predators (Kummer, 1971). In contrast, there is another more 'defensive' form, characteristically shown by a small animal cornered by a larger predator. In such circumstances, a strategy with the elements of surprise and unpredictability would be advantageous: tonic immobility, followed by sudden escape, is shown by many birds when caught by a predator (Archer, 1979). The jump-bite of the rat (Leyhusen, 1973), although very different in form, shares the same broad function, to maximise the chance of escape when the animal has effectively been trapped by a predator.

The jump-bite combines surprise with the infliction of pain. Unlike the more offensive forms of antipredator aggression, it occurs only after escape has been tried and has failed. It is therefore clearly fear-motivated in terms of both form and causation. For these reasons, it has generally been termed 'defensive' (Adams, 1979, 1980; R. J. & D. C. Blanchard, 1981; D. C. & R. J. Blanchard, 1984a; see section 2.3 in relation to Brain's classification).

The occurrence and distribution of a similar characteristic form of exclusively fear-motivated aggression in species other than the rat is not well documented. Adams (1980) has argued that it occurs in many species of muroid rodents, and D. C. & R. J. Blanchard (1984a) have argued that it is widespread in mammals, occurring in cats, bears and even in humans. It would seem beyond dispute that fear motivation can influence the form of aggressive behaviour in the direction of more defensive components in a wide variety of vertebrate species, and that fear-motivated aggression occurs when danger is accompanied by blocked escape. However, it is likely that characteristic and stereotypic forms of fear-motivated defensive aggression, such as the jump-bite, have evolved only in small animals which are likely to be trapped by a larger predator. Even then, such responses would not have evolved as a matter of course: they would have had to fit in with the general antipredator strategy of that species, which would have evolved in relation to its ecology, predation pressure and fighting capability relative to that of a predator.

Other examples of characteristic 'defensive' antipredator aggression, comparable to the jump-bite of the rat, include a similar reaction by hamsters (Leyhausen, 1973) and lemmings, *Lemmus lemmus* (Arvola *et al.*, 1962), possibly the 'lunge-bite' shown by a defensive polecat in intraspecific encounters (Poole, 1974), and the kicks directed by a cornered rabbit to a stoat (Edmunds, 1974). Adams (1980) has, in addition, suggested that the counter-attacks observed in the intraspecific fights of voles (Clarke, 1956) and other muroid rodents, and the avoidance leaps of *Peromyscus* spp. (Eisenberg, 1962) are comparable to the rat's jump-bite. Nevertheless, it would again seem useful to make the distinction between the *capability* of showing defensive or fear-motivated aggression (which is probably widespread) and the occurrence of a distinctive, stereotypic and readily emitted antipredator response such as the jump-bite (which is probably restricted to a relatively few species). However, any conclusions reached on the basis of present studies should be regarded as tentative as this subject has not been systematically investigated.

Predation pressure may lead to selection for *decreased* protective aggression in some instances. Huntingford (1976*b*) suggested that in species where predation pressure is high but no effective antipredator attack is possible, there will be selection for alternative antipredator defences (Edmunds, 1974; Archer, 1979). If, in such a species, intraspecific fighting is conspicuous and cannot be brought to an end quickly, Huntingford argued that selection will favour low levels of both protective and competitive aggression. She cited the deermouse *Peromyscus maniculatus bairdii* as one example, and guppies when subject to high predation pressure as a second.

From the discussion in this section, it is possible to identify verbally some of the likely evolutionary rules which could account for protective aggression. First, it will occur when there is a high degree of potential predation on an animal which is capable of driving off a predator, either by itself or in a group; secondly, it will occur in an animal which is likely to be cornered by a predator, in which case, a more defensive or fear-motivated form of aggression is likely to arise. Finally, high predation pressure may lead to low levels of protective aggression in some circumstances.

3.2 Similarities and differences between protective and competitive aggression

Although predation can be viewed as the major selective agent accounting for individual differences in protective forms of aggression, their occurrence in any particular species is not restricted to responding to a predator. They also occur in response to circumstances which share similar eliciting stimuli to the presence of a predator, for example a sudden lunge from another animal, or a sharp pain. Rapid attack in response to sudden pain probably provides the mechanism behind the counter-attacks observed in several species (e.g. Bernstein & Gordon, 1974; Geist, 1974*a*, *b*; Jakobsson *et al.*, 1979), as well as the pain-induced aggression found in the laboratory. In functional terms, such counter-attacks conform to a strategy identified in game theory analyses (Chapter 9) called 'retaliator' (Maynard Smith, 1982). Formal models show that in a population of animals which tend either to attack indiscriminately or never to attack, individuals which only attack when first attacked would form an evolutionarily stable strategy. Thus, counter-attack towards a conspecific by an otherwise relatively non-aggressive animal is likely to have evolved in many circumstances.

In some animals, antipredator protective aggression is similar to competitive (intraspecific) aggression. Huntingford (1976*b*) reviewed studies showing that the black-headed gull, the ten-spined stickleback and some other species show similar forms of behaviour to an aggressive conspecific and to a predator. The same weapons may be also used in antipredator and competitive fighting (Edmunds, 1974; Huntingford, 1976*b*).

Antipredator and competitive aggression may also co-vary with reproductive condition (in sticklebacks and baboons: Huntingford, 1976*b*) or with gonadal hormone levels (in doves: Huntingford, 1976*b*). In baboons, they show similar sex differences. Huntingford (1976*a*) observed the same individual three-spined sticklebacks' responses to intra- and interspecific territorial intruders and to a potential predator. She found that the level of aggression in individual fish not only co-varied across different types of territorial intruders but also correlated with behavioural reactions to the predator which were indicative of 'boldness', i.e. lack of fear and willingness to approach.

There is, therefore, evidence that in some species, protective aggression is similar in form and causally related to competitive (intraspecific) aggression. In other species, this is not the case: in these, there will still be causal links between antipredator and other forms of protective aggression, but the links do not extend to competitive forms of aggression. In a review of the social behaviour of ungulates, Geist (1974*a*) suggested that although damaging counter-attacks have primarily evolved for antipredator defence, they may also be used in intraspecific contests. However, natural selection will have operated to reduce the occurrence of damaging attacks in intraspecific fights. In the Norway rat, R. J. & D. C. Blanchard (1981) found that aggression by a colony intruder in response to a dominant colony male was similar in form to pain-induced and antipredator aggression; however, all three were distinct in form from the offensive attack of a dominant male in its own colony.

Where there are such differences between protective and competitive aggression in the same species, competitive aggression is observed to be the less damaging form. The usual functional explanation for this is as follows. First, antipredator aggression (which determines the form of other types of protective aggression) is most effective if it is as damaging as possible: this accounts for the evolution of damaging counter-attacks in ungulates (Geist, 1974a) and baboons (Kummer, 1971), and the more wide-ranging target-sites in the defensive compared to the offensive aggression of the Norway rat (R. J. & D. C. Blanchard, 1981). Secondly, where antipredator aggression has particularly damaging consequences, selection pressure will have minimised the potential to cause such damage in intraspecific fights (as in the above example of the Norway rat, and a comparable restriction of biting sites found in the rhesus monkey: Bernstein & Gordon, 1974). Geist (1966, 1978) pointed out that fighting is costly to an animal with weapons, largely because both combatants are likely to be damaged through retaliation (see also Geist, 1974b). More formal game theory models demonstrating selection against damaging contests, and towards ritualised displays and threats, were advanced by Maynard Smith & Price (1973), and are discussed in Chapter 9.

It is now apparent that in a wide range of species, intraspecific contests are fought with different and less damaging weapons and motor patterns than those directed towards a predator, and that a principle such as the danger from retaliation can account for the origin of many of these cases. Such species include the giraffe, the moose (Edmunds, 1974), the oryx, the marine iguana, and rattlesnakes (Eibl-Eibesfeldt, 1961; Huntingford, 1976b).

R. J. & D. C. Blanchard (1981) have suggested an alternative way in which intraspecific aggression has become less damaging than antipredator aggression for the rat. Attack directed by a resident towards an intruder results in 85–90 per cent of the bites being directed to its dorsal surface. This occurs despite the fact that the intruder's behavioural strategies (e.g. lying on its back and boxing) serve to reduce the access of the resident to these areas. Blanchard & Blanchard suggested that such a restriction of attack to the less vulnerable dorsal areas has resulted from kin selection (Maynard Smith, 1964; see also Clutton-Brock & Harvey, 1976, for a discussion of the role of kin selection in reducing aggression). They pointed out that the young males from a colony are attacked by the adult males, thus tending to drive them out. But since they are likely to be related to the adult males, it will be advantageous for these males not to harm them (by killing or castrating them, as could happen if the abdominal region were the target of bites). A few bites directed to the back are sufficient to drive them from the immediate area. In support of this hypothesis, Blanchard & Blanchard reported individual variability in the degree to which bites are limited and they observed one rat which did not limit its bites to the back at all. This resulted in the death of all other males in its own colony as well as the death of all intruders. They concluded that it is unlikely that such an individual's male relatives would have survived to reproduce.

One possible parallel with the separation of intraspecific offensive and protective forms of attack is the separation of different antipredator responses in some species of birds. Kruuk (1964) found that black-headed gulls distinguished between different predators with regard to the frequency with which they attacked them, ranging from the repeated attacks (and attacks over a wide area) on a fox, to the low frequency attacks directed towards a hedgehog.

Curio (1975) has reviewed a number of examples of where different classes of predator produce different protective responses. In his own study of the pied flycatcher, he found that whereas snarling attack would be directed at predators who were a threat to the nestlings but would not be capable of killing the adults, more dangerous species, which were predators of the adults and fledged young, were mobbed from a safe distance. These two distinct antipredator responses represent a causal separation of parental aggression (considered in Chapters 5 and 6) and more generalised protective aggression. Curio regarded the latter as a compromise between the requirements for self-preservation and brood-protection (discussed further in Chapter 5).

In this section, I have argued that where antipredator aggression is damaging in its effects (and damage to predators will be selected for), there will be selection against using this form of aggression in intraspecific fights. A possible parallel to this is found in the specialisation of antipredator responses to different predators.

3.3 Sex differences

There is no obvious evolutionary reason why there should be widespread sex differences in protective aggression. As Milligan, Powell & Borasio (1973) have argued, mechanisms for defence against predation would, in general, be equally adaptive for males and females.

Indeed, in species where antipredator aggression is found, it generally occurs to a similar extent in both sexes (e.g. see Edmunds, 1974; Huntingford, 1976b): this excludes parental forms of protection, where there is often specialisation by one sex (see Chapter 5).

On the basis of the discussion of interspecific differences, we should expect any exceptions to this generalisation to involve one or both of two features: sex differences in the capability of driving off a predator, and sex differences in predation pressure.

The first is relevant to species where one sex is much larger and more readily shows aggressive responses than the other, as a result of sexual selection (see Chapter 7). Consequently, if predation pressure is pronounced, the more aggressive sex would be more effective in predator defence, and hence would come to play an exclusive or disproportionate role in group defence. For example, in baboons, the much larger males are often more aggressive to predators than are the females (Kummer, 1971).

The second possibility is more tenuous: if the two sexes have been subject to different predation pressure, owing to their occupation of different ecological niches, this might produce differences in protective aggression. A difference in the home ranges of males and females has been reported for a variety of birds (Selander, 1972) and mammals (e.g. European voles: Frank, 1957; mountain goats: Geist, 1978; and primates such as the orang-utan: Wrangham, 1980); it has also been argued that intrasexual competition is generally infrequent in mammals (Hrdy, 1981; Floody, 1983). Nevertheless, there have been few attempts to relate such differences to different degrees of predation pressure which might have produced sex differences in antipredator aggression.

One exception to this is the discussion by Geist (1978) of the home ranges of male and female mountain goats. He suggested that males congregate away from females to exploit peripheral areas. In such a habitat, grouping becomes an adaptive antipredator strategy, and there would be selection against damaging weapons. If we link this discussion to the present argument, it is apparent that we have identified conditions which could, in theory, lead (or have led) to sex differences in antipredator aggression.

There is, however, a wide gap between such a general functional explanation and the results of the many studies of laboratory rodents showing behavioural sex differences, in pain-induced fighting (Conner *et al.*, 1983) and in characteristics associated with emotional responsiveness (Archer, 1971, 1975; Gray, 1971, 1979; Beatty, 1979). Gray (1971) suggested that male rodents show greater aggressiveness and fearfulness (but see Archer, 1971, 1975, and Floody, 1983). The original functional explanation for this putative difference, put forward by Gray & Buffery (1971), was based on the group-selectionist argument of Wynne Edwards (1962), which is now generally regarded as incorrect (see Chapter 1). It also relied upon a narrow and misleading view of mammalian social organisation restricted to considering male dominance (cf. Wrangham, 1980; Hrdy, 1981; Floody, 1983). In addition, the issue of whether it is appropriate to characterise sex differences on behavioural measures used in laboratory tests (e.g. of forced exposure to a novel area, emergence from the home cage, active avoidance and passive avoidance) in terms of a single characteristic is still problematic (e.g. see Archer, 1979; Beatty, 1979; Suarez & Gallup, 1981; van Oyen, 1981).

Nevertheless, in view of the many findings of sex differences in laboratory studies of rodents, it seems worth while examining briefly whether there are any functional grounds for supposing that males and females could be subject to different selection pressures for characteristics associated with defensive aggression. There is, in fact, some evidence from older naturalistic and semi-naturalistic studies of the Norway rat for intrasexual segregation, or at least for little competition between the sexes (Steiniger, 1950; Calhoun, 1963), and this is borne out by more recent laboratory studies showing that males attack male intruders and females attack females (e.g. D. C. Blanchard *et al.*, 1984*b*; Debold & Miczek, 1984). Nevertheless, it seems doubtful whether such a behavioural segregation of the sexes would be drastic enough to produce different antipredator responses (see also Milligan *et al.*, 1973; Conner *et al.*, 1983).

Are there any other ways in which the sexes might have come to differ in their degree of fear-motivated behaviour? One possibility, related to the issue of intrasexual behavioural segregation, is raised by the study of van der Poll *et al.* (1982). They found that defeats in aggressive encounters produced relatively permanent behavioural changes in male rats (consisting of avoidance of the opponent's side of the cage, reduced locomotion and weight loss) which were not observed in females. These effects were found to be androgen-dependent. Although the extent of the behavioural changes must have been exacerbated by continued confinement in a small cage with a victorious opponent, they do, nevertheless, indicate a sex difference which could have a functional basis in restricting the potential damage involved in losing a fight. For a male rat, seeking to prolong a fight against a victorious opponent, or seeking to re-start such a fight, would be disadvantageous in view of the possible damage that might be inflicted. Fights between females tend to be less damaging (D. C. Blanchard *et al.*, 1984; DeBold & Miczek, 1984) and hence the need for avoidance of a victor tends to be less important.

This argument provides an extension of that of R. J. & D. C. Blanchard (1981), that defensive behaviour serves to restrict the amount of damage incurred in a fight. Avoidance of a victor after defeat carries further these attempts to minimise damage during the fight.

Another reason why males and females could come to differ in a characteristic related to defensive aggression was suggested by Giles & Huntingford (1984). In a study of antipredator responses shown by sticklebacks at seven different sites in Scotland, they found that males were generally 'bolder' than females, i.e. showed a stronger tendency to approach a predator, and would flee less readily than females. Giles & Huntingford suggested that after an encounter with a predator, the males would need to return to their reproductive territories more quickly than would the females (Giles, 1984), because it is the males which form territories and subsequently defend the young (see Chapters 5 and 8). Although this suggestion is purely speculative, and is *post hoc* rather than predictive, it could be tested by comparing sex differences in antipredator aggression in teleost species with different patterns of parental care. We should also note that the explanation links the sex difference in aggression with the part the male plays in parental care, and, consequently, it would apply only to species where male territoriality is associated with subsequent parental defence (Chapter 5).

In this section, we have considered several possible reasons why males and females might show differences in defensive aggression. They are associated with sex differences in antipredator responses, or the need to restrict the damaging consequences of fighting, or reproductive and parental behaviour. Each one would be expected to produce sex differences, but only in those species where the particular selective pressure had operated. This more cautious conclusion contrasts with suggestions of a general sex difference of the sort advocated by Gray & Buffery (1971).

3.4 Life cycle changes

Early in life, practically all animals are vulnerable to predation, and in many species, one or both parents provide antipredator protection at this time (Chapter 5). The point at which there is a change-over from parental protection to the appearance of the offspring's own antipredator responses will, of course, vary in different animals, notably between altricial and precocial species. In many reptiles, for example, protective aggression is found shortly after hatching (Burghardt, 1978).

We should, in general, expect there to be two conflicting selection pressures: the first is for the parents to show increased defence as the offspring become older (the reasons for this – in terms of the increasing cost of replacing the offspring – are discussed in Chapter 5); the second, opposing pressure is for the parents to show decreased defence when the offspring are able to defend themselves.

Trivers (1974) put forward the 'parent–offspring conflict' model to account for the evolution of characteristic features in the relationships between parents and their offspring, and between different offspring of the same parents. It was based on the conflict of interest between parents and offspring, and between individual offspring, which is predicted from the pattern of relatedness in sexually reproducing species. For example, an offspring can achieve twice as much benefit in terms of fitness by obtaining a resource for itself rather than letting a sibling obtain the same resource (since the sibling will be related by a half to the offspring we are considering). Similarly, if there were no other considerations, a parent would benefit twice as much by obtaining a resource for itself than by aiding an offspring (which contains only half of that parent's genetic material). However, since the offspring are younger, more helpless and (usually) more numerous than the parent, the latter would generally gain more in terms of fitness by increasing the offspring's survival chances.

Trivers' model predicts that there will come a time during an offspring's development when the parent or parents will benefit more by switching their parental care (or 'investment') from that particular offspring either to younger existing ones or to renewed reproductive investment in future offspring. Because of the pattern of relatedness outlined above, this switch in investment would occur before it would be beneficial to the offspring we are considering. (Later on, there will come a point when it will be beneficial to both parent and offspring: the older offspring would be receiving so little benefit from continued parental care that this would be outweighed by the indirect benefit to fitness obtained from the parent's producing more offspring.)

The important point for the present discussion is that offspring will tend to benefit from receiving parental care for longer than parents will benefit from giving it. A commonly cited example of this principle is weaning conflict (Trivers, 1974; Dawkins, 1976; Duncan, Harvey & Wells, 1984). However, it should also apply to protective aggression. It would therefore be interesting to assess, for example, whether mammalian young start defending themselves only once the parents have ceased doing so: in particular, are they responsive to signals from the parents that parental aggression is declining? Or do they use their own protective responses while their parents are still defending them? Posing the question as two alternatives probably represents an oversimplification, in that an offspring which was able to defend itself against localised attack would possess an advantage over one which relied solely on parental defence. Such additional benefits would complicate any predictions made solely on the basis of the parent–offspring conflict model.

There are very few developmental studies with which to assess these functional speculations. Evidence from reptiles, reviewed by Burghardt (1978), does show that in this group of animals, where there is relatively little parental protection, protective aggression (and other forms of antipredator behaviour) occur shortly after birth in many species. Some, such as pit vipers, will stand their ground and strike. In the snakes generally, some show antipredator behaviour shortly after birth, whereas, in others, it is days or weeks before it appears.

Some birds show antipredator aggression early in life. At the beginning of this chapter, I mentioned the bluff displays of young owl chicks (Edmunds, 1974). In mammals, Owings & Loughry (1985) noted that pups from a feral population of prairie dogs would attack bullsnakes but not the more dangerous rattlesnakes (which *are* attacked by adults). Ebert (1983) observed that wild housemice would show biting in response to tactile stimulation (the experimenter's hand, or forceps) when they were as young as 12–13 days, and could draw blood from human handlers at the age of 17 days. The appearance of protective aggression at this time during development approximately coincides with the decline in maternal aggression in this species (Svare, 1983). Similarly, in *Peromyscus* (*leucopus* and *maniculatus*), Wolff (1985) found that 17- to 18-day-old pups would counter-attack adult conspecifics (which are infanticidal towards younger pups but will generally show little or no aggression to pups of this age).

In contrast to these observations on *Mus* and *Peromyscus*, a cross-sectional study of shock-induced fighting in male laboratory rats (Hutchinson, Ulrich & Azrin, 1965) found that 24-day-old rats showed practically no fighting and 33-day-old animals very little. Thereafter, the probability of fighting increased steadily with older age groups until 90 days, the oldest age group used. The rats used in this study were housed in single-sex groups, thus allowing the opportunity for social interactions. It was also found that social isolation from 22 to 90 days of age reduced but did not eliminate shock-induced fighting.

These findings on the development of protective aggression are little more than what Bekoff (1981) referred to as 'isolated bits and pieces of information held together by rather soluble glue'. He went on to point out that we lack longitudinal developmental studies of aggressive behaviour which take account of ecological context, and which therefore could be used to assess functional hypotheses.

3.5 Future research directions

Although Brain (1981*b*) and others have identified protective aggression (or self-defence behaviour) as a coherent functional group, it has rarely been studied or discussed in these terms. Rather, the evidence reviewed in this chapter comes from a series of individual research topics, which, when described together, may appear fragmentary and unrelated. Antipredator aggression has often been considered in relation to other forms of antipredator behaviour (Edmunds, 1974; Cloudsley-Thompson, 1980) but rarely in connection with other forms of aggression (Huntingford, 1976*b*, is a notable exception). Brain's own research on protective aggression was concerned with hormonal influences, rather than functional aspects. Similarly, the extensive laboratory research on pain-induced aggression involves few functional hypotheses. Where it has (e.g. in relation to sex differences), they have been based on a restricted perspective provided by the study of a few species in a single experimental set-up. Functional hypotheses of any generality are unlikely to be generated from such a source. We also saw in the previous section that the developmental evidence consisted of fragments which were difficult to place in a wider framework.

In view of these considerations, future research on functional aspects of protective aggression needs to begin by constructing a broad framework rather than filling in the gaps of previous work. Existing studies provide a number of verbal functional hypotheses to account for the occurrence of interspecific and other variations in protective aggression. These are based mainly on its antipredator function. One future task would be to extend existing descriptive studies to provide a systematic and comprehensive description of antipredator behaviour in a wider range of animal groups, and to investigate its relationship to other forms of protective aggression. Here we are following the basic ethological ground-rules, described by Tinbergen (1963), concerning the importance of a descriptive phase in the study of behaviour.

Such descriptive studies would enable *informed* functional hypotheses to be formulated (again, see Tinbergen, 1963), and eventually to be tested experimentally. Specific functional hypotheses would initially be concerned with the survival consequences of different forms of protective aggression. An example can be provided from the Norway rat, which has been well studied already. Several researchers have suggested that the function of the jump-bite is to escape from a predator when cornered. It should be possible to compare the effectiveness of animals which show jump-bites and those which do not in escaping from different predators in different circumstances. Such studies would have to be carefully designed to take account of ethical guidelines covering experiments on predation and aggression (Huntingford, 1984b).

The function of antipredator aggression is to remove the predator. How this is achieved can vary. In many cases, the animal engages in potentially damaging forms of attack, whereas in others extensive bluff is involved (section 3.1). A further task for future research is to provide a functional explanation for the evolution of antipredator bluff. In the case of competitive aggression, game theory models predict that animals will evolve the capacity to distinguish bluff from a real signal under most conditions (Maynard Smith & Parker, 1976); the exception occurs when there is a possibility of serious injury resulting from a mistake (Maynard Smith & Parker, 1976). Presumably the same would apply for antipredator aggression.

The formal game theory approach to function can also provide a useful parallel direction to any detailed descriptive work and to verbal functional hypotheses. Formal models can be used to pose the questions derived from descriptive studies in more general terms. Such an interconnecting parallel approach of formal models and field studies has already proved successful in relation to competitive aggression (e.g. Austad, 1983; Clutton-Brock, Guinness & Albon, 1982: see Chapter 9).

Finally, as Bekoff (1981) suggested, a broad descriptive approach which takes account of the animal's ecological context can be extended to cover longitudinal developmental studies of aggression, and these can again be assessed in relation to functional hypotheses.

Further reading

Edmunds, M. (1974). *Defence in Animals: a survey of anti-predator defences*. Harlow: Longman.
Geist, V. (1974). On the relationship of social evolution and ecology in ungulates. *American Zoologist*, **14**, 205–20.

Geist, V. (1978). On weapons, combat and ecology. In *Advances in the Study of Communication and Affect, vol. 4: Aggression, Dominance and Individual Spacing*, ed. L. Krames, P. Pliner & T. Alloway, pp. 1–30. New York: Plenum.

Huntingford, F. (1976). The relationship between inter- and intra-specific aggression. *Animal Behaviour*, **24**, 485–97.

4

Mechanisms of protective aggression and their relation to function

4.1 General principles

In this chapter, we consider the mechanisms by which protective aggression can achieve its function. In Chapter 1, three possible ways in which function might be related to mechanism were outlined. In relation to protective aggression, these can be illustrated by considering the design of a device whose function is to react to disturbance by delivering a noxious stimulus. The exact effector system is immaterial: it might be a loud noise (e.g. a burglar alarm) or a physically harmful device (e.g. a booby-trap).

In the simplest version of such a mechanism, there are two sets of design features which will determine how it operates. One is what starts if off and the other is what stops it. There are two broad ways of starting it off: the first is that a specific stimulus has to appear, and the second is that a representation is made of the expected features of the outside world and the input has to differ from this.

The first can, in principle, operate very simply, for example where any tactile stimulus or nearby movement sets the mechanism off: this would be very indiscriminate and would correspond in the animal world to the sea anemone aggression described in Chapter 2. It could also be likened to a burglar alarm which is activated by the detection of movement. A more discriminating version would involve a stimulus configuration containing one or more specific features which would trigger (or unlock) the mechanism: this would correspond to the classical ethologist's description of specific stimuli ('social releasers') which were regarded as operating a 'releasing mechanism' for particular types of behaviour. Here, a matching principle starts the mechanism off. An alternative would involve a desired end-point being represented as a reference value, which is compared with the current input. In its

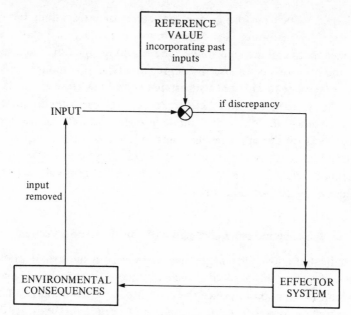

Fig. 4.1. Control system model of simple form of protective aggression.

simplest form, the desired end-point could be a stored value representing previous input signals, and a mismatch starts the mechanism off.

There are two ways of ending the effector mechanism's action. The simplest is for it to operate for a particular period of time (possibly determined by physiological variables such as fatigue or time taken to discharge the noxious stimulus). Such a device is found in sea anemone aggression. More usually, the consequences of the effector action determine its cessation, i.e. when the specific input that started the mechanisms off is no longer perceived (in the case of matching mechanisms) or when there is no longer a discrepant input (in the case of a mismatch mechanism). In the second case, the behavioural mechanism can be depicted as a control system incorporating a negative feedback loop. Fig. 4.1 shows how this would operate for a simple form of protective aggression. More complex versions would involve either complex representations of the expected (or desired) input, or several alternative outputs, operating on the basis of additional internal or external influences.

So far, we have considered protective mechanisms with relatively simple design features. In principle, more complex systems can be envisaged, which perform a constant series of integrations between a wide variety of inputs, assessing a number of current environmental circumstances and internal states. The resulting signals could affect not only the occurrence but also the degree and type of response, and could make moment-to-moment adjustments to the response. If such principles were incorporated into a physical device, it would represent a great advance in terms of complexity and subtlety of responding, compared with the other mechanisms we have considered. It would be achieved most readily in a computer-operated system. However, it will be argued that such complex mechanisms are not normally required for protective aggression (section 4.7).

4.2 General principles applied to protective aggression

As indicated above, the aggressive responses of anthozoan coelenterates operate on the basis of a relatively indiscriminate input-matching mechanism. In addition, once the response has been initiated, it carries on in a stereotyped form until completed (Brace & Pavey, 1978). At present, there are few examples of this sort, but we should expect a similar mechanism to be widespread in invertebrates with simpler nervous systems.

In vertebrates, there is one form of protective aggression which has been described as 'reflexive' in nature. This is pain-induced aggression. Ulrich & Azrin (1962) argued that it is a reflexive response to the painful stimulus, on the grounds that it persists over repeated trials, and gradually increases over repeated sessions. Both of these contrast with the habituation of aggressive responses which is typical in other cases (e.g. territorial aggression: Archer, 1976a).

There is, however, some doubt as to the correctness of Ulrich & Azrin's interpretation. Crosby & Cahoon (1973) argued that since most studies have employed consistent lengths of shock intensity, superstitious learning related to shock-offset could account for the continuation and persistence of pain-induced aggression. In other words, the rats were being reinforced for attack by the offset of shock. The evidence for this view included Crosby & Cahoon's own observations that when rats were given shocks of a variable duration (so as to remove the link between the response and shock offset), attack declined over subsequent trials. In addition, when low-intensity shocks were administered at a high frequency to squirrel monkeys, habituation occurred (Hutchinson, Renfrew & Young, 1971). Thirdly, a *gradual* increase in the level of pain (as opposed to the sudden pain of an electric shock) was found not to elicit attack even though the same *level* of pain, administered suddenly, did so (Ulrich, 1966).

Although the finding that rats will maintain shock-induced fighting for approximately 15 000 responses over a period of 7.5 hours (Ulrich & Azrin, 1962) did seem to indicate an automatic mechanism involving little modification as a result of the consequences of behaviour, it does seem more likely that such persistence of responding represents superstitious avoidance learning. If this view is correct, the mechanism underlying pain-induced aggression would be better described by a simple version of the control system shown in Fig. 4.1. In this case, the reference value would be based simply on the absence of a painful input (representing in functional terms an absence of immediate damage to the body surface). This reference value is compared against the input, and errors beyond a certain value will activate the receptor system, inducing first escape and, if this is blocked, attack, until the pain-inducing stimulus has been removed.

This control system could show a limited amount of habituation, i.e. the reference value would be capable of some modification as a consequence of a sustained painful input. But such modification is unlikely to extend too far, since pain provides a signal that bodily tissue is likely to be damaged.

Pain-induced aggression, therefore, provides an animal with a very simple mechanism for showing an immediate response to sudden physical danger which, under natural conditions, usually results from an attack by another animal or potential predator.

In the case of a specific sensation such as pain, the difference between describing the mechanism in terms of a reaction to mismatch, rather than a matching mechanism triggered off by a specific stimulus may be largely semantic. However, when we turn to considering antipredator reactions in vertebrates, the issue of whether the animal is responding to a specific stimulus (a match) or to novelty (mismatch) becomes more salient.

It has been argued by Bolles (1970) and by Curio (1975) – researchers from two very different traditions (North American animal learning and German ethology, respectively) – that animals show species-specific defence reactions which are innately biased to respond to particular stimulus configurations normally associated with specific predators. The maladaptive nature of having to rely on associative learning to escape from a dangerous predator was pointed out independently by both researchers.

Nevertheless, observational learning (e.g. through the alarm calls from conspecifics), and an innate response to general classes of stimuli (particularly those involving novelty) could go a long way towards explaining the specificity of antipredator responding, particularly in social mammals. But there still remains evidence that some of these responses are innate, in the sense that a particular response is given to a small number of specific stimuli, normally associated with only one type of predator, and these specific connections cannot readily be accounted for in terms of past experience with that predator.

Curio (1975) has argued that this applies to the antipredator attack of the pied flycatcher, since the mobbing responses develop early in life irrespective of experience with the effective stimuli (owls or shrikes). He constructed a model of the antipredator responses of the pied flycatcher, involving specific matching mechanisms. Curio used the Lorenzian concept of an innate releasing mechanism (IRM), a mechanism characteristic of the species or sex, set to respond to particular stimulus configurations. (He defends the cautious use of the IRM concept against criticisms such as those of Hinde, 1970.) Several specific configurations are regarded as activating IRMs for mobbing, whereas others activate IRMs for snarling attack (the latter being concerned with protecting the nestlings and the former with more general antipredator defence: see Chapter 3). Curio also included an IRM for novel birds: however, responses to novelty would be better described by the mismatch mechanisms referred to above.

It is interesting to note that Spanish pied flycatchers do not show the mobbing response to the red-backed shrike, which is absent from this part of the flycatchers' range, thus demonstrating that the precise content of the matching mechanisms reflects the adaptive requirements of the environment.

4.3 A discrepancy model of aggression

In Curio's study, novel birds were attacked. It has often been noted in general discussions of aggression that novel stimuli evoke attack (e.g. Hebb, 1946; Archer, 1976a; Marler, 1976). 'Novel stimulation' is a very general concept which can be applied to any case where an animal possesses the neurosensory equipment to store representations of past experience, and to compare these representations with the current input. If this comparison process were coupled with an effector system which removes the animal from the input or delivers a noxious stimulus, it would provide the animal with the ability to respond to potential as well as to currently damaging events. In other words, responding to *any*

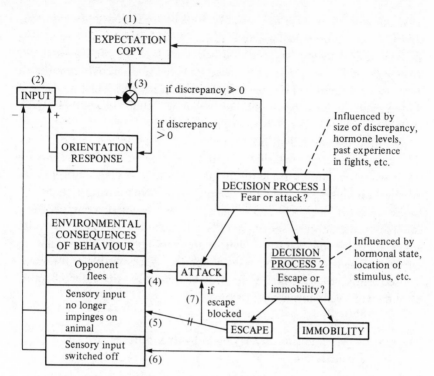

Fig. 4.2. A simplified version of a control theory model of aggression and fear in vertebrates proposed by Archer (1976*a*). (From Huntingford (1984*a*), p. 82.)

major change in the immediate surroundings would provide a simple way of responding to potential danger. Such a novelty-activated mechanism (shown earlier in Fig. 4.1) could also provide the basis for attack or avoidance following individual distance intrusion. If the defended area were extended to include the surrounding location, a possible mechanism for territorial behaviour would result.

Fig. 4.2 shows a motivational model of aggression and fear, incorporating the principles of mismatch detection and negative feedback already shown in Fig. 4.1. This was originally proposed as a general model of aggression and fear in vertebrates (Archer, 1976*a*). Although a number of reservations have been made about the generality of this model (e.g. Huntingford, 1980, 1984*a*; Toates, 1980), it does provide a convenient representation of cases where novelty and other forms of mismatch evoke aggression. Fig. 4.2 is a simplified version of the original diagram (Archer, 1976*a*), taken from Huntingford (1984*a*). It shows that a representation of the familiar environment, or expectation copy (1), is compared (3) with the input (2) and that any discrepancy above a certain level will activate one of two behavioural effector systems, attack, or fear (escape–avoidance) responses, whose consequences are to remove the discrepant input or to remove the animal from it. Several internal and external factors, as well as the magnitude of the original discrepancy, influence whether attack or fear responses occur (Decision Process 1).

As Huntingford (1984*a*) noted, this model does not explain precisely the sequence of actions during a fight. More generally, it cannot explain cases where responses are carefully matched to very specific properties of the stimulus (e.g. how the opponent behaves during a fight) or assessments of the costs and benefits of the specific encounter. Such complex integration of specific inputs will, as indicated earlier, require more complex mechanisms, which are probably less important in protective aggression than in cases of fighting over a particular resource; consequently, this general topic will be left until the later chapters dealing with competitive aggression.

4.4 Mechanisms underlying individual differences in responsiveness

We now consider the mechanisms which could account for some of the species and sex differences in protective aggression, discussed in Chapter 3.

In the motivational model described above, the same general input is capable of evoking either aggression or fear, depending on variables such as the degree of discrepancy, hormonal levels, and past experience in fights. In discussions of the motivation of aggression, it has been widely recognised that similar situations lead to aggression or fear behaviour or both (e.g. Hebb, 1946; Hinde, 1956; Tinbergen, 1957; Baerends, 1975; Archer, 1976*a*).

One important variable in accounting for interspecific or intersex differences might be the level of input at which attack is replaced by fear. This point can be illustrated and developed in relation to the discrepancy model outlined above (Fig. 4.2: Archer, 1976*a*). I suggested that as the discrepancy magnitude increased, the response tendencies for both attack and fear increased, but the tendency for fear rose more rapidly than that for aggression (an assumption originally derived from Berkowitz, 1962: Archer, 1976*a*). Fig. 4.3 illustrates this relationship graphically. Fig. 4.4 shows the relationship between discrepancy magnitude and the probability of fear behaviour (F) and attack (A). The

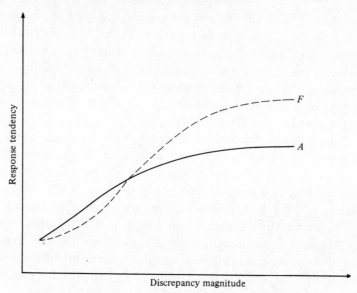

Fig. 4.3. Graph showing hypothetical relationship between discrepancy magnitude and tendency to attack (A) or show fear behaviour (F). In this and subsequent graphs, the term 'tendency' is used in the way described by Hinde (1970), to denote that causal factors for this type of behaviour are present, but that the response may not occur when these factors are too weak or are inhibited by another tendency. (From Archer (1976a), Figure 3, p. 259.)

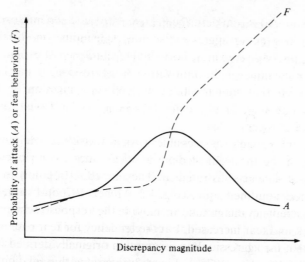

Fig. 4.4 Graph showing hypothetical relationship between discrepancy magnitude and the probability of attack or fear behaviour (obtained by combining Figures 4 and 5 from Archer (1976a)).

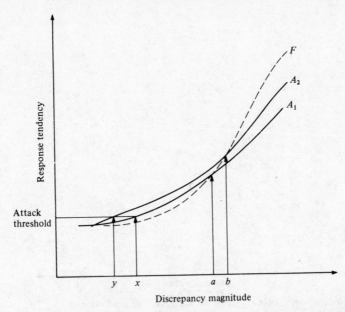

Fig. 4.5. Graph showing the influence of a variable which increases the response tendency of attack from A_1 to A_2. This results in the point where F overtakes A being moved from a to b, i.e. a higher discrepancy magnitude. It would also lead to a lowered attack threshold, i.e. moved from x to y.

decline in the probability of attack at higher discrepancy magnitudes represents the suppressive effects of the rising fear tendency on attack. This incorporates the notion that once the tendency for fear has overtaken that for attack, it is difficult for attack responses to replace fear responses (an idea that has been depicted in catastrophe theory representations of aggression and fear: Zeeman, 1976; Toates, 1980; Colgan, Nowell & Stokes, 1981).

In principle, there are three possible ways in which the relationship shown in Fig. 4.3 can be changed by variables such as phylogenetic history, hormones or past experience. The first one (Fig. 4.5) is that there is an increase in the response tendency for attack, but not in that for fear. This results in the attack threshold being lowered and the fear threshold increased. (This example corresponds to the influence of androgens, described below.) The second possibility (Fig. 4.6) is that the response tendency for fear, but not for attack, is increased. This will lead to a lowering of the threshold for fear, but will have no effect on that for attack. In practice, this will be difficult to distinguish from cases

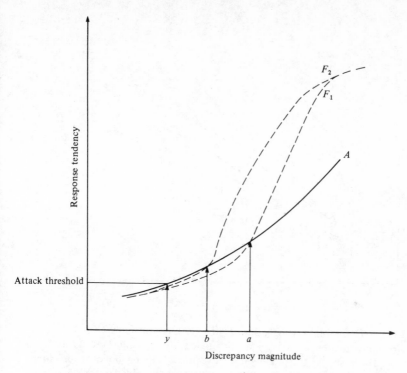

Fig. 4.6. Graph showing the influence of a variable which increases the response tendency of fear from F_1 to F_2. (It is assumed that this accentuation of fear tendency will not apply at low discrepancy magnitudes, so that there remain some circumstances when the animal will still respond aggressively, but we could also envisage a totally fearful animal where the curve for F_x always exceeded that for A.) This results in the point where F overtakes A being moved from a to b, i.e. being lowered. The threshold for attack (y) remains unchanged in this case.

where there is also a decrease in the response tendency for attack (Fig. 4.9). The third possibility (Fig. 4.7) is that the tendencies for both attack and fear are increased to an equal degree: this results in a lower threshold for both attack and fear. One possible example of this is the short-term influence of corticosterone, discussed in Chapter 8. (Logically, there are two further possibilities, shown in Figs 4.8 and 4.9: first, an increase in attack combined with a decrease in fear, and secondly, an increase in fear combined with a decrease in attack. In practice, these will be difficult to separate from Figs 4.5 and 4.6.)

We now consider how these principles may apply to selection for the relative readiness to show aggression or fear. In Chapter 3, we described

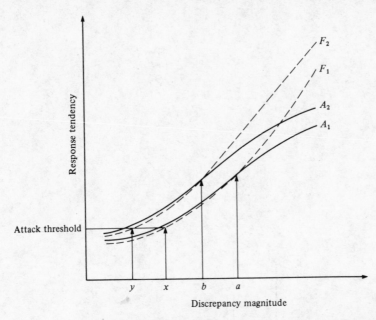

Fig. 4.7. Graph showing influence of a variable which increases the response tendencies of both attack and fear from A_1 and F_1 to A_2 and F_2. This results in the point where F overtakes A being moved from a to b, i.e. a lower discrepancy magnitude. It would also lead to a lowered threshold for aggressive responses, i.e. moved from x to y.

the finding that sticklebacks which more readily attacked a predator or territorial intruder also showed indications of lack of fear – or 'boldness' – in situations not involving attack (Huntingford, 1976a;). These results suggest that there may have been selection for a decrease in fear tendency (e.g. from F_2 to F_1 in Fig. 4.6), which would indirectly lower attack by raising the threshold for fear responses (from b to a in Fig. 4.6). Of course, it is possible that an increase in the response tendency for attack might also be involved (resulting in a movement from F_2 to F_1 and from A_2 to A_1 in Fig. 4.9), but there is no evidence that specifically indicates this. A similar change could also account for the greater 'boldness' of a male than a female stickleback when approaching a predator (Huntingford, 1976a).

Where there is selection for low levels of both protective and competitive aggression, owing to a combination of predation pressure, an absence of effective antipredator attack, and the conspicuousness of

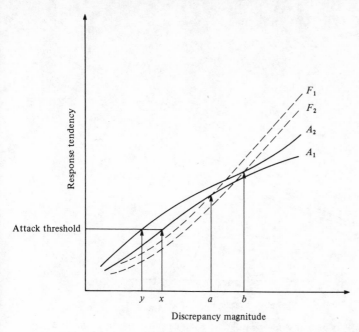

Fig. 4.8. Graph showing the influence of a variable that increases the response tendency of attack from A_1 to A_2 and decreases the response tendency of fear from F_1 to F_2. This results in a decrease in the threshold for attack from x to y, and the point where F overtakes A being moved from a to b. Overall, the direction of the effect on behaviour would be the same as that shown in Fig. 4.5, where there was no lowering of fear, but the level of discrepancy where this animal would still show attack would be higher.

intraspecific fights (Huntingford, 1976*b*), a lowering of the threshold where fear replaces aggression could have occurred. Such a response pattern may have resulted from selection in the opposite direction to that which produced Huntingford's 'bold' stickleback (represented by a move from a to b in Fig. 4.6).

Tinbergen (1957) argued that selection on territorial aggression favours neither the tendency to attack nor the tendency to escape in isolation from one another, but operates on their relative strengths. A similar consideration is important for the above discussion: a way of varying behavioural responsiveness is achieved by altering the *balance* of fear and attack tendencies shown to a single variable representing an input signal.

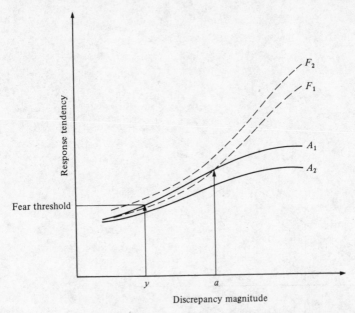

Fig. 4.9. Graph showing the influence of a variable that increases the response tendency of fear from F_1 to F_2 and decreases the response tendency of attack from A_1 to A_2. This results in a lowering of the threshold for fear responses, so that they are never prevented from occurring by attack. They cease at the lower threshold for fear, i.e. when the discrepancy would not normally activate either attack or fear. It was noted in Fig. 4.6 that this could also produce a non-aggressive, fearful animal: the only way to distinguish these cases is by examining their aggressive tendencies after the influence of a fear-reducing procedure.

4.5 Mechanisms underlying the separation of protective and competitive aggression

So far, we have considered mechanisms producing general differences in responsiveness. As indicated in Chapter 3, some species have evolved very different forms of aggression to serve protective and competitive functions. I argued there that the basic form of aggression is that used in response to potential danger, such as a potential predator, or a painful stimulus, and that this has been further modified by selection operating in the context of competitive (intraspecific) aggression to reduce its damaging effects (Geist, 1966, 1978; R. J. & D. C. Blanchard, 1981).

In representing the causal mechanisms involved, the first consideration is that there are two forms of effector system, each presumably under the control of different causal factors. Since it would be important for each system to respond only to the appropriate set of circumstances, we should expect selection to have produced different decision processes for the two forms of aggression. There are at least two possible ways in which this could have occurred.

The first applies to cases where there is complete separation of protective and competitive fighting. Different weapons and different forms of behaviour are involved. Protective attack is shown in response to stimuli such as sudden pain or a suddenly approaching novel stimulus, and also learned stimuli which signal the likelihood of a predator. The simple feedback mechanism described in the first section of this chapter would describe the basic mechanisms. However, in the case of competitive fighting, mechanisms involving more detailed and specific assessment of the input, and which influence a different effector system, would be required. In addition, once the aggressive interaction has begun, many other variables will influence the animals' responses to each other's behavioural signals and selection will have operated on the sensitivity and flexibility of responses shown in an aggressive exchange, to produce a range of fighting strategies (Chapters 9 and 10).

Such a complete separation of the mechanisms used for protective and competitive aggression will have occurred in species where it is important to use different effector systems and weapons in the two sorts of fighting. However, an alternative way of producing two different reactions, from the same underlying motivational mechanisms, could operate in species where antipredator behaviour is fear-motivated and there is less potential for damage through specialised weapons. In the case of the rat – and to a lesser extent in other small mammals – it has been argued on neurophysiological and behavioural grounds that protective forms of aggression are under the control of fear motivation, and are used only when escape is blocked (e.g. R. J. & D. C. Blanchard, 1977, 1981; Adams, 1979; see Chapters 2 and 3). Before outlining a possible mechanism for these cases, we consider the behavioural evidence for the motivational separation of protective and competitive aggression in rats and mice.

R. J. & D. C. Blanchard (1981) have made detailed comparisons between shock-induced fighting and aggression towards a colony intruder in rats. They found that attack by a dominant male towards a stranger in a colony shows a number of characteristic features. First, it is preceded by approaching and sniffing the stranger (Fig. 4.10a). If the smell of an adult male is perceived, piloerection occurs, and this is shortly followed by biting and chasing. The intruder rears on its hind-

(a) *Left*: Sniff.
   Right: Elevated crouch.

(b) *Left*: Defensive upright posture.
   Right: Offensive sideways posture.

(c) *Left*: Aggressive posture.
   Right: Submissive posture.

(d) *Left*: Offensive upright posture.
   Right: Defensive upright posture.

(e) Mouse. *Left*: Offensive upright posture.
   Right: Defensive upright posture.

Fig. 4.10. Acts and postures used in fighting by rats (*a–d*) and mice (*e*). (From Grant & Mackintosh (1963), pp. 249, 252 and 254.)

legs, facing its attacker and the attacker makes offensive sideways movements towards the attacker's back (Fig. 4.10*b*). The intruder shows mostly defensive postures such as rolling over to lie on its back (Fig. 4.10*c*). This is effective in reducing biting because the attacker does not bite the ventral surface or the legs (Chapter 3). Detailed observations confirmed the impression that this was a strategy to produce relief from biting by reducing access to preferred biting sites rather than a submissive signal which restricted the attacker's bites.

If a dominant rat is removed from its colony and given foot shock in the presence of another rat, it will show the typical defensive reactions characteristic of a colony intruder. Its acts and postures will be very different from those used to attack an intruder, consisting mainly of the defensive upright posture ('boxing'), shown in Fig. 4.10*d*. The target

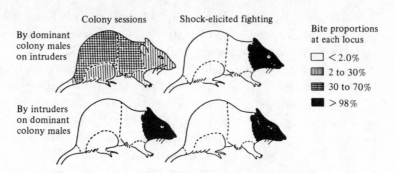

Fig. 4.11. Location of bites on different parts of the opponent's body, comparing offensive attack by a dominant rat (top left) with defensive attack by a colony intruder (bottom left), and with shock-induced fighting for the same animals (right-hand side). (From Blanchard & Blanchard (1981), Fig. 2.)

sites are also different, being largely restricted to the snout instead of the back (Fig. 4.11).

R. J. & D. C. Blanchard (1977, 1981) also noted that the responses of any shocked rat were very similar to the defensive postures shown by an intruder when attacked by a dominant. R. J. Blanchard *et al.* (1984*b*) showed that whereas a fear-inducing stimulus (a cat) abolished offensive attack by a colony member to an intruder, it did not affect defensive aggression in response to shock.

From this evidence, it was concluded that shock-induced aggression is, like defensive aggression when attacked, a fear-induced response occurring when escape is blocked (see Chapter 3).

Studies of shock-induced aggression in laboratory housemice have produced less clear-cut results. Brain, Al-Malki & Benton (1981) have argued that it shows a more offensive form. They found that manipulations such as castration, food deprivation, anosmia, dexamethasone and ACTH influenced shock-induced aggression in ways that paralleled effects on social conflict. Although only a restricted behavioural analysis was carried out in this study, the researchers indicated that a wider range of postures, including social investigation and submission, was apparently shown than for the rat. There was also a strong tendency for the mice to fight in the inter-shock interval, which was not characteristic of shock-induced fighting in rats (R. J. & D. C. Blanchard, 1981), and indeed was one of the reasons why it was originally referred to as reflexive fighting. Brain *et al.* concluded that shock-induced attack shows more of an offensive nature in the mouse than in the rat, and that any differences from socially induced fighting, such as the higher proportion of ventral bites, were a consequence of the upright postures of the mice when responding to shock.

The Blanchards have remained unconvinced by this argument. R. J. Blanchard (1984) argued that earlier studies of shock-induced aggression in mice (e.g. Legrand & Fielder, 1973) did not demonstrate an *offensive* component in the upright 'boxing' posture, and nor did Brain *et al.*'s study. Blanchard's research group set out to replicate Brain *et al.*'s procedure with a detailed behavioural analysis (described by R. J. Blanchard, 1984). They found no increase in offensive behaviour in socially naive mice tested with anosmic opponents under conditions of shock (Brain *et al.*'s procedure). Offensive behaviour was, in fact, low in both shocked and unshocked groups and showed no increase even after repeated testing. Another experiment showed that repeated shocks in a confined chamber dramatically suppressed the offensive behaviour shown by a territorial mouse, and even a few intense shocks would suppress offensive attack for several minutes. It would seem from these results that earlier studies on mice did not distinguish between offensive and defensive forms of the upright posture (shown for the mouse in Fig. 4.10*e*).

Blanchard's results do seem to indicate that the form of shock-induced attack is defensive in the housemouse as well as in the rat, and support the argument that it is fear-motivated and similar to antipredator aggression.

One difficulty with this interpretation is the nature of hormonal influences on shock-induced attack. Brain *et al.* based their argument that shock-induced attack is offensive in the mouse largely on the similarities between hormonal and other influences on this form of attack and on offensive attack. Conner *et al.* (1983) have also argued that the facilitative effects of androgens on shock-induced fighting suggest that it has an offensive component to it. We return to a possible explanation for these results after considering the possible mechanisms producing the separation of offensive and defensive attack.

The two forms of attack may result from the same general class of inputs, but occur at different intensities of stimulation. In terms of the discrepancy model described earlier, defensive aggression would occur in response to the higher discrepancy magnitudes which typically evoke fear responses. In the original model (Archer, 1976*a*), there was a direct link from blocked escape to attack, which was shown as indistinguishable from offensive forms of attack (Fig. 4.2). In view of the clear distinction between offensive and defensive (fear-motivated) aggression, it is now necessary to modify the model to show a third form of

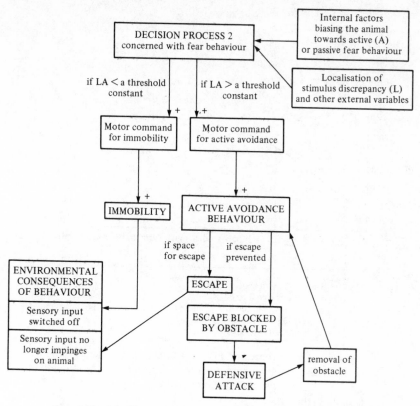

Fig. 4.12. Modification of the original discrepancy model (Archer, 1976a), to take account of fear-motivated 'defensive attack'.

fear behaviour occurring when active escape is blocked. Fig. 4.12 shows part of the original model (Archer, 1976a, Fig. 1) modified to take account of this.

A possible mechanism, based on the discrepancy model, which could account for both types of attack is as follows. First let us suppose that the decision process for protective forms of aggression involves simple internal representations which are readily and rapidly compared with the input (as in the case of pain, discussed earlier), and the point at which the response tendency for fear exceeds that for attack is readily reached in these circumstances (Fig. 4.13). In contrast, when a dominant colony rat responds to a novel male intruder, the signals entering the decision process are more complex, involving both an assessment of the discrepant nature of the input and an analysis of its specific properties (male odour), as indicated by the behavioural description outlined above. In this case, a more finely tuned set of

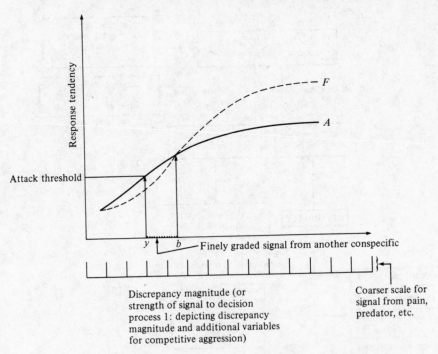

Fig. 4.13. Graph (based on Fig. 4.3) depicting the relationship between discrepancy magnitude (or more accurately the strength of the signal to decision process 1) and response tendency for attack (*A*) and fear (*F*) in the rat.

signals (*y–b* in Fig. 4.13) are sent to the decision process. (Of course, other variables, such as a history of successful fights, androgen levels, etc., will also act to bias the decision in favour of attack.)

It is therefore envisaged that the finely graded signal resulting from investigating a novel male conspecific is more likely to fall in the lower region (*y–b*), which will typically result in offensive attack. In contrast, the more coarsely graded signal likely to result from pain or the presence of a predator will, in most cases, overshoot the narrow range which produces offensive attack, and result in fear responses, i.e. defensive attack if escape is blocked.

Incidentally, once offensive and defensive attack are viewed in terms of competing response tendencies, it becomes possible to explain the influence of androgens on shock-induced attack as arising from an increase in the (sub-threshold) offensive tendency exerting a facilitating effect on the largely defensive shock-induced attack.

4.6 Developmental flexibility

According to the classification of adaptive mechanisms introduced in Chapter 1, developmental flexibility provides a second way in which an animal can change its behavioural mechanisms so as to fit the adaptive requirements of the environment. We turn now from characteristics which are typical of a species or one sex to consider the ways in which behavioural development is affected by different rearing environments. There are, however, few studies investigating the influence of developmental manipulations on protective forms of aggression. Hutchinson *et al.* (1965) and Hutzell & Knutson (1972) found that rats which were reared in social isolation after weaning showed less shock-induced fighting than those which had been reared socially. Hutzell & Knutson found that shock-elicited *biting* was not affected by isolation rearing. It seems likely that the reduction in other components of shock-induced fighting was associated with the absence of opportunities to engage in any type of aggressive interaction from weaning onwards. A more recent study found that male rats reared with castrates from weaning to adulthood showed less shock-induced fighting than males reared with intact cagemates (Knutson & Viken, 1984). Observation of behaviour in the home cage revealed a reduced rate of aggression between the intact rats and the castrates.

Although these studies provide only isolated examples of developmental influences, they indicate that the maintenance of defensive forms of agonistic responding is dependent on experiences of aggressive encounters during young adulthood. The scarcity of the evidence again reinforces Bekoff's (1981) appeal for more ecologically and evolutionarily oriented approaches to behavioural development (see Chapter 3).

4.7 Short-term flexibility

The third way in which animals can change their behavioural responses according to the adaptive requirements of the environment is through short-term mechanisms, principally learning. In addition, hormonal changes resulting from a social experience such as winning or losing a fight (Leshner, 1975) can change behavioural responsiveness in adaptive ways over short periods of time.

It is known that defeat in agonistic encounters produces hormonal changes, notably lowered androgen levels and increased adrenocortical secretion (in the swordtail: Hannes, Franck & Liemann, 1984; the mouse: Leshner, 1975, 1980, 1983*b*; the rat: Schuurman, 1980; the rhesus monkey: Rose, Gordon & Bernstein, 1972). Whether these hormonal changes affect protective aggression is not clear. In the rat and the mouse, androgen levels seem to affect offensive attack but not submissiveness or avoidance (Leshner & Moyer, 1975; Schuurman, 1980). There is, however, evidence that castration does lower shock-induced aggression in the male rat (section 4.5), so that defeat may have an inhibiting effect on shock-induced attack, mediated through a lowering of testosterone levels.

The evidence on whether pituitary–adrenocortical hormones affect shock-induced aggression is not clear. Brain (1981a) concluded that 'acutely-injected ACTH sometimes . . . augments and sometimes suppresses electroshock-induced fighting in the laboratory rat'. Which one occurs depends on the dose. The evidence for ACTH influences in mice is similarly unclear (Brain *et al.*, 1981). Brain (1981a) concluded that relatively high levels of glucocorticoids increase shock-induced aggression in laboratory rats and mice; if this conclusion is reliable, and if we are dealing with physiological levels, it would be consistent with the suggestion that fear-responding is enhanced by glucocorticoid secretion (Leshner, Moyer & Walker, 1975; but see Leshner, 1981, 1983b), provided that shock-induced aggression is viewed as completely fear-induced. Alternatively it would also be consistent with increased adrenocortical activity enhancing aggressiveness (Heller, 1977, 1978, 1979), if we regard shock-induced aggression as being influenced by such a component (section 4.5; and Conner *et al.*, 1983).

Protective aggression can also be modified by learning. Shock-induced fighting in rats has been classically conditioned to a buzzer and to a tone (Creer *et al.*, 1966; Vernon & Ulrich, 1966). In addition, discrimination and reversal learning have also been demonstrated (Lyon & Ozolins, 1970). These studies, although limited, do indicate that defensive attack can be conditioned so that the animal can respond to new situations in an adaptive way.

Perhaps a more surprising result is that of Azrin, Hutchinson & McLaughlin (1965b), who showed that squirrel monkeys would learn, in response to shock, to pull a chain to produce an inanimate object which they then attacked. On the face of it, this result suggests that the biting response to shock has an offensive component to it, since the reinforcing properties of offensive attack have been shown in a number of studies (e.g. Lagerspetz, 1964; Legrand, 1970; Tellegen & Horn, 1972) whereas fear motivation is aversive (D. C. and R. J. Blanchard, 1984b). Alternatively, a two-stage process could be involved whereby fear-motivation is aroused by the shock and then reduced by the biting response, thus providing an equivalent to shock-avoidance learning. Fear reduction may be an important component in reinforcing and maintaining pain-induced attack generally.

Follick & Knutson (1978) have demonstrated a suppression of shock-induced aggression by subsequent shock in the rat. Presumably, there was a shift from one form of fear response, which was punished, to another, which was not.

In later chapters (Chapters 7 to 10), we consider short-term flexibility in competitive aggression, where it is important for the animal not only to respond to an assessment of the resources and the costs of fighting for them, but also to assess the likelihood of its success, and how these might change throughout the fight. In contrast, antipredator aggression (and defensive aggression in general) would appear to benefit much less from detailed assessment strategies (except perhaps when the predator species is numerous and is responded to only when it shows behavioural signals indicating behavioural intent: see Kruuk, 1972). Antipredator responses have to be made quickly, and it is likely that a brief assessment of target properties, such as size, and speed and type of movement, would provide sufficient information for making a rapid and appropriate response. Kruuk (1964), in a study of the antipredator behaviour of the black-headed gull, remarked that the gulls often reacted to the behaviour of the predator rather than its morphological characteristics since the time taken to recognise a species may be maladaptive. Of course, we have already noted that many prey species show different antipredator responses to different predators (e.g. Kruuk, 1964; Curio, 1975) but even in such cases there is no indication that a detailed assessment, of the sort involved with resource competition, is undertaken.

4.8 Conclusions, suggestions for future research and ethical issues

In this chapter, we first considered general principles underlying the mechanisms of protective aggression. We then discussed how changes in the relative motivational strengths of aggression and fear could alter behavioural responsiveness and form the basis of differences in responsiveness between species and between the sexes. The causal separation of protective and competitive aggression was regarded as being achieved either by the complete separation of the motivational mechanisms for the two forms, or alternatively by distinguishing them as offensive or defensive (fear-motivated) attack. The limited studies of developmental flexibility indicated that social experience in aggressive encounters was necessary for the full expression of shock-induced aggression in rats. Short-term flexibility was discussed in terms of a possible role for the hormonal consequences of defeat in decreasing shock-induced attack. Studies of learning showed classical and instrumental conditioning of shock-induced fighting in rats, and it was suggested that fear reduction was important in instrumental conditioning. It was argued that protective aggression is unlikely to involve complex assessment strategies.

A number of general issues were raised in this chapter, indicating broad areas for future research. For example, studies of invertebrates such as sea anemones indicated reflexive mechanisms little influenced by the consequences of behaviour. A more systematic investigation of these mechanisms would provide one interesting avenue for future research. The relative importance of stimulus matching and mismatch mechanisms on the input side was another general issue which was highlighted, in relation to vertebrate antipredator aggression.

Much of the present experimental evidence concerning mechanisms of protective aggression is derived from shock-induced aggression in laboratory rodents, mainly rats. Although it has been fairly clearly established as a defensive form of aggression, several questions remain unanswered, in particular, whether the influence of castration and replacement treatment indicates that there is also an offensive component (Conner *et al.*, 1983), and whether the persistence of shock-induced aggression results from superstitious learning (Crosby & Cahoon, 1973).

Although such questions still remain unanswered, I would argue that alternative ways of investigating protective aggression need to be sought in the future, for two reasons: first, on ethical grounds, and secondly, because of the restricted nature of the evidence obtained from this procedure.

We have already noted in passing in Chapter 1 that research on shock-induced aggression is ethically questionable, and in this chapter we referred to Ulrich & Azrin's (1962) infamous experiment in which they subjected pairs of rats to approximately 15000 shocks over a period of 7.5 hours, until 'the rats were damp with perspiration and appeared to be weakened physically' (p. 514). Originally, the justification for using shock-induced aggression was that it provided a convenient 'model' of aggression, i.e. a simple procedure that would readily and reliably evoke aggressive responses. This was regarded as important for studying the precise stimulus conditions under which aggression occurred

(Ulrich, 1966) and, it was hoped, for providing some information that could generalise to human aggression (Ulrich *et al.*, 1973; Berkowitz, 1983, 1984). In addition, shock-induced aggression has been used as a convenient behavioural procedure in fields such as physiological, psychology and behavioural pharmacology (Rodgers, 1981*a*). However, the research of the Blanchards and others, discussed earlier, which shows that shock-induced aggression is not typical of offensive or 'angry' aggression, undermines its usefulness as a general model of aggression. In addition, many of the original studies were of an atheoretical nature, guided by a Skinnerian tradition of measuring differences in a standard form of behaviour following systematic variations in stimulus conditions. Very few general principles have been derived from such studies.

It is now widely agreed that ethical assessments of animal experimentation must take into account the *value* of the research as well as the suffering involved (e.g. Archer, 1986*b*; Bateson, 1986). Shock-induced aggression can be regarded as a procedure entailing an unusually high level of suffering for a low level of scientific interest (Drewett & Kani, 1981), and for this reason alone it is not recommended as a suitable direction for future research.

The repeated use of shock-induced aggression to investigate systematic variations produced by changing the stimulus conditions and the internal physiological state can also be criticised on the grounds that it has restricted psychological and physiological investigations in terms of both the species used and the topics studied (cf. Beach, 1950).

If we were looking for the most useful single study on mechanisms underlying protective forms of aggression, we would not find it among the shock-induced aggression studies. Instead, we might turn to the study of Huntingford (1976*a*), who compared fear responses with aggressive reactions to a predator and a territorial intruder, in individual sticklebacks. Or we might look to the work of the Blanchards, who investigated defensive aggression in its appropriate social context. Future advances in our general understanding of the mechanisms underlying protective aggression in different species are more likely to be made by following up these studies than by further use of the shock-induced aggression procedure.

Whatever technique or line of research is chosen, ethical questions will still be involved in any future studies of protective aggression, since it is, by definition, aggression directed towards a possible threat to the animal. Defensive aggression such as that shown by the rat can be generated only by a fear-inducing stimulus such as a potential predator, pain or a threatening or attacking conspecific. If fear cannot be eliminated from such studies, at least physical damage can. The requirement *not* to subject animals to repeated and prolonged attack, whether from a conspecific or from a predator, is just as important as the need to protect them from repeated electric shocks. Huntingford (1984*b*) has discussed ethical issues surrounding research on predation and aggression and made a number of recommendations for reducing suffering in such cases. As well as restricting or eliminating physical attacks, she suggested using model predators. These might profitably be used in future studies of protective aggression, as Huntingford has in her own investigations of the antipredator reactions of sticklebacks to a model pike.

In addition to studies aimed at understanding the general principles underlying mechanisms of protective aggression, other aspects which require empirical study for a wide range of species include the relationship between antipredator (protective) aggression and competitive aggression (e.g. territorial) on the one hand, and fear responding on the other. Systematic studies of how such characteristics change in relation to selective breeding, the developmental environment and learning influences would show how the mechanisms respond to environmental change in the long, medium and short term.

Further reading

Archer, J. (1976*a*). The organization of aggression and fear in vertebrates. In *Perspectives in Ethology 2*, ed. P. P. G. Bateson & P. Klopfer, pp. 231–98. New York & London: Plenum.

Blanchard, R. J. & Blanchard, D. C. (1981). The organization and modeling of animal aggression. In *The Biology of Aggression*, ed. P. F. Brain & D. Benton, pp. 529–61. Rockville, Maryland: Sijthoff & Noordhoff.

Curio, E. (1975). The functional organization of anti-predator behaviour in the Red Flycatcher: a study of avian visual perception. *Animal Behaviour*, **23**, 1–115.

Ulrich, R. E. (1966). Pain as a cause of aggression. *American Zoologist*, **6**, 643–62.

5

Parental aggression: a functional analysis

In many species, parental and protective aggression overlap, in that similar antipredator attack will be shown whether or not eggs or young are present. There are, however, a number of further issues involved with parental aggression. First, predation directed towards the young is a more widespread phenomenon than predation towards adults, since eggs and young provide a particularly vulnerable phase in any animal's life history. Many animals will seek to eat the eggs or young of a species whose adults they would not approach. Secondly, cannibalism of the eggs or young, as well as infanticide serving other functions, are widespread in the animal kingdom (Ridley, 1978; Hrdy, 1979; Labov *et al.*, 1985). Thirdly, the parents are not simply defending their own lives. Therefore, the degree of risk they are prepared to take will be related to variables such as the numbers of offspring and the cost of replacing them. These and other factors combine to make a functional analysis of parental aggression more complex than for the forms of protective aggression considered so far.

In Chapter 3, it was suggested that predation pressure, combined with the ability to drive off a predator, would lead to the evolution of protective aggression. Animals whose nests and young are likely to be subject to predation will be even more likely to evolve antipredator aggression, since alternative strategies such as escape are less likely to be available where eggs and young are also at risk. Nevertheless, some alternative antipredator strategies involving the young are found, albeit in a more limited form than in the case of a single adult confronted with a predator. These include carrying the young (e.g. mouth-brooding and cutaneous incubation in fish), and camouflaging or hiding them (e.g. caching in ungulates). In other cases (e.g. birds such as the ostrich), the young are sufficiently precocial to show effective escape or avoidance responses of their own early in life (Wilson, 1975, chapter 16).

Nevertheless, active aggression towards predators or conspecifics (or both) is usually the only strategy available to parents protecting the young. In many species, group defence occurs, and high predation pressure may have led to the evolution of colonial nesting and group defence in birds (Horn, 1968).

In this chapter, we first briefly survey the distribution of parental aggression in the animal kingdom, and then consider some empirical studies designed to assess how effective parental aggression is in preventing predation or infanticide. We then consider a formal model, that of Andersson *et al.* (1980), which might account for interspecific differences in parental aggression, and look at some relevant empirical studies. This is followed by a brief discussion of the importance of infanticide in the evolution of parental aggression. In relation to sex differences, we consider two types of formal model, lifetime optimality and game theory, which concern the evolution of patterns of sex differences in parental care in different taxonomic groups. It is argued that these general models of parental care patterns are relevant to the distribution of parental aggression between the two sexes in different animals. Finally, we return to the model of Andersson *et al.*, to consider changes in parental aggression with the age of the parent.

5.1 The distribution of parental aggression in the animal kingdom

In the majority of metazoan species, fertilisation is external and developing eggs receive little attention (Eisenberg, 1981). Formal models of the function of parental care usually emphasise the increased survival prospects that it produces (e.g. Maynard Smith, 1977, Andersson *et al.*, 1980), particularly in otherwise unsafe or variable environments (see also Tallamy, 1984).

Parental care occurs in a variety of invertebrate groups. Defence of eggs or young has been observed in polychaetes, arachnids (Ridley, 1978; Saito, 1986*a*, *b*), crustaceans (Montgomery & Caldwell, 1984), and in insect groups such as Hemiptera (e.g. Tallamy, 1982), Isoptera, Hymenoptera (Tallamy, 1984) and Dictyoptera (e.g. Breed, Hinkle & Bell, 1975). Discussing the evolution of parental care in insects, Tallamy (1984) concluded that 'In its most primitive state, parental care is limited to the physical protection of eggs.' Protection extends beyond the egg stage in a smaller number of insect species, for example in neotropical tortoise beetles, some sawflies, lace bugs and treehoppers of the family Membracidae (Tallamy, 1984). While in the majority of cases it is the females which defend the eggs or young, there are species where both sexes contribute. In some cases, such as African reduviids of the genus *Rhinocoris* and (possibly) the giant water bug *Lethocerus americanus*, the male defends the eggs (Wilson, 1971). In these species, mechanisms for ensuring paternity are found (Tallamy, 1984). Saito (1986*a*) has also observed biparental or male parental defence in an arachnid, the spider mite (*Schizotetranychus celarius*), which lives gregariously in a nest built in bamboo, and attacks the larvae of the phytoseiid spider predator *Typhlodromus bambusae*. Since the sex ratio is female biased and there is male haploidy, Saito suggested that there has been kin selection for paternal defence in this species.

Among fish, egg-guarding occurs in lungfish such as *Protopterus*, and in many teleosts. Examples of male, female and biparental care are found. Defence by the male is commonly a by-product of territorial behaviour, the male having built a nest and guarded the surrounding area prior to fertilisation. There is therefore an overlap between territorial aggression (Chapter 7) and parental aggression in teleosts such as the blenny (*Pholis*) and the stickleback (Eisenberg, 1981; Ridley, 1978).

Egg-guarding is also found in a number of species of Amphibia. In some salamanders, for example the male is territorial and drives away conspecifics (including adult females) which would otherwise eat the eggs (Ridley, 1978); in other species, egg-guarding is carried out by the female (e.g. Forester, 1979). Protecting the eggs against predators is found in a minority of anurans (Wells, 1981): in some species of Leptodactylidae (South American frogs), the males are territorial and defend the eggs (Ridley, 1978). In one species from this group, aggressive behaviour such as biting and wrestling has been observed in defence of the eggs against conspecifics (Townsend *et al.*, 1984).

Among reptiles, parental care is highly developed in the crocodile, where the female guards the nest. In the Squamata (lizards and snakes), egg-guarding also occurs in some species: for instance, the female cobra (*Nija*, sp.) coils above the clutch of eggs, and in some lizards egg-guarding has been reported. However, there are few systematic or detailed observations of aggressive responses by the parent or parents guarding the eggs.

Among birds, where parental care involves more complex activities such as incubation of the eggs and feeding the young, egg and brood defence are widespread. Studies have been carried out on the black-headed gull (Kruuk, 1964), the eider (Ahlen & Andersson, 1970), the arctic tern (Lemmetyinen, 1972), the fieldfare (Andersson *et al.*, 1980), and many other species.

In mammals, parental care (including aggression) is commonly restricted to the female, owing to lactation, although there are species in which the male defends the young (e.g. the arctic ground squirrel: McLean, 1983). Parental aggression by female mammals has generally been referred to as 'maternal aggression' (Moyer, 1968; Svare, 1981), and it has been observed under natural conditions in species from a variety of orders: for example, elephant seals (McCann, 1982), squirrels (Sherman, 1981), hedgehogs (Burton, 1969), zebra, eland and wildebeest (Kruuk, 1972), baboons, rabbits, sheep, cats and moose (Svare, 1981), and several rodent species (Ostermeyer, 1983).

5.2 Studies of the effectiveness of parental aggression

The first type of functional evidence we examine comes from studies assessing the effectiveness of parental aggression in terms of the increased survival rates it may produce in the offspring. These studies have involved both those species where the main threat to the young arises from predators and those where it comes from infanticidal conspecifics.

Most of the evidence is derived from field studies. In some cases, these provide descriptive data. Forester (1979) observed that parental care by the female led to decreased predation in the mountain dusky salamander (*Desmognathus ochtophaeus*). In a study of fieldfare, Andersson *et al.* (1980) found that where nest defence was high, fewer chicks were later taken by predators than where nest defence was low. Blancher & Robertson (1982) found that pairs of Eastern kingbirds whose nests were not taken by predators were more aggressive than those whose nests were preyed upon. McLean (1983) also found some evidence that paternal defence was associated with lowered mortality rates in an arctic ground squirrel.

Other field studies involve experimental manipulations. Tallamy & Denno (1981) found that when nymphs of the lace bug (*Gargaphia solani*) were deprived of the antipredator defence of their mother, in an environment containing predators, significantly fewer survived to maturity. Similar findings were reported for the green lynx spider, *Peucetia viridans*, by Fink (1986), and Townsend *et al.* (1984) found that fewer eggs were cannibalised as a result of defence by the male in a neotropical frog. When lactating Belding's ground squirrels were restrained so that they could not defend their territories, the rate of infanticide rose sharply (Sherman, 1981).

Saito (1986*b*) found that the success of the spider mite *S. celarius* in killing predator larvae depended on the numbers of females and males in the nest. The males were more effective than females in killing the larvae, and the presence of the parents in the nest considerably enhanced the numbers of predator larvae which were killed.

Other experimental evidence is provided by laboratory studies of rodents. Mallory & Brooks (1978) showed that reduced attacks by the female collared lemming (*Dicrostonyx groenlandicus*) towards a strange male led to the male killing more pups. However, Erskine, Denenberg & Goldman (1978) found that aggressive lactating female rats did not prevent males from entering the nest nor did they behave differently towards males which killed pups and those which did not. Hahn (1983) found similar results for mice. Consequently, they questioned whether post-partum aggression does serve to protect the young in these species. Nevertheless, there are several problems with this conclusion: first, naturalistic studies may be a more valid way of testing the effectiveness of parental aggression (Svare, 1981); secondly, inbred laboratory strains of rodents may have been selected for low levels of maternal aggression (Ebert, 1983); and, thirdly, in these and other studies the intruder could not escape from the vicinity of the nest, and may have killed the pups for this reason. To overcome this last problem, Wolff (1985) used two interconnecting cages. He found that maternal aggression by sympatric species of *Peromyscus* (*P. leucopus* and *P. maniculatus*) did effectively ensure a high level of pup survival. In the absence of the mother, the neonates were nearly always killed or attacked by a strange adult.

Other studies demonstrate that high-ranking or highly aggressive lactating females have a higher rate of offspring survival than less aggressive females (Svare, 1981) and there are similar findings from field studies of elephant seals (Reiter, Panken & Le Boeuf, 1981; McCann, 1982). Whether these results are attributable to the consequences of maternal aggression or to reproductive suppression in low-ranking females (cf. Hrdy, 1981; Floody, 1983) remains to be established.

5.3 A functional model of parental aggression

The studies reviewed in the previous section assessed the survival rates of the offspring in the presence or absence of parental aggression. They all involved assessments of very simple verbal hypotheses in the tradition of the classical ethological approach to function. A different, and more sophisticated, approach to studying the survival value of parental aggression is by the use of formal models. At present, there is one very general model which seeks to predict the occurrence and level of parental aggression in a wide range of circumstances. This model, proposed by Andersson *et al.* (1980), will be discussed in some detail, since it enables widespread and systematic predictions to be made. The model incorporates the costs and benefits of parental aggression, and seeks to predict the optimal level of parental defence in different circumstances. More precisely, it aims to establish the optimal individual strategy for lifetime reproductive success. It can therefore be applied to topics such as differences between species, and changes throughout the breeding season and across the life-span (section 5.6).

In general terms, it is obvious that the costs of parental aggression are the risks of injury or death, and the benefits are the continued survival of the offspring (and the parent in many cases). However, both the risks and the benefits will vary according to circumstances and it is these relationships which are quantified in the model. It is assumed that the risks will vary with the particular type of defence used: the most effective strategy will be a rapid direct attack on the predator, but this would also be the most costly; threats or alarm calls from a distance are less risky, but also less effective. It is therefore assumed that a range of different levels of parental defence is possible, and as the effectiveness in deterring the predator increases, so do the costs.

A second variable incorporated into the model is the number of offspring: the more offspring that are being defended, the greater will be the benefit in terms of fitness (i.e. the representation of the parent's genes in the next generation). Thirdly, the probability that the parent

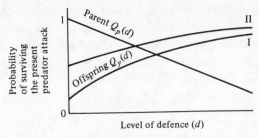

Fig. 5.1. Probability of surviving a predator attack upon offspring for parent, $Q_p(d)$, and for offspring $Q_y(d)$ (I and II), when the parent uses antipredator defence of level d. With offspring curve II, the probability of escaping predation when parents provide no defence ($d = 0$) is higher than it is with curve I. (From Andersson *et al.* (1980), p. 537, Fig. 2, original caption.)

will survive until the next reproductive season, and the probability that the offspring will do so, are both represented in the model. The parents' survival chances are always greater than the offspring's, but both the absolute and relative differences decrease with age. Finally, each level of defence employed by the parent is assumed to have an associated probability of parent survival, and a probability of offspring survival.

A formula incorporating these variables was derived to give the optimal level of defence at different offspring and parental survival probabilities and for different brood sizes. It also predicts changes over the breeding season and the life span (section 5.6). One crucial variable is the degree of danger the offspring are under in the absence of parental aggression, which Andersson *et al.* represented graphically as the probability of surviving the present predator attack (Fig. 5.1). If this is high even in the absence of parental aggression, then a standard amount of parental aggression will produce only a small increase in fitness. If the probability of surviving the predator attack is low in the absence of parental aggression, a standard amount of parental aggression will produce a much larger increase in fitness. The study of Tallamy & Denno (1981) on the lace bug (see previous section) illustrates the way in which predation pressure can influence the effectiveness of maternal defence. Fig. 5.2 shows that nymph survival is increased (sevenfold) by

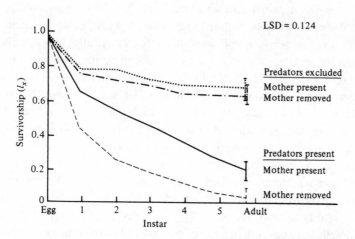

Fig. 5.2. Field comparisons of the effect of maternal care on nymphal survivorship in the lace bug *Gargaphia solani* in the presence ($N = 26$) and absence ($N = 24$) of predators. LSD = Least Significant Difference multiple comparison test. Means differing by this interval or more are significantly different ($P < 0.05$). (From Tallamy (1984), originally in Tallamy & Denno (1981), original caption.)

maternal defence when predators are present, from the low level found in unprotected eggs. In contrast, where there are no predators, maternal defence has little effect on the already high survival level.

The model also predicts a higher level of parental aggression when there are more young in the brood or litter. This has been demonstrated in species of mammals, birds and fish. Svare (1981) reported that, in mice, a higher overall level of maternal aggression was associated with larger numbers of young in the litter. Robertson & Biermen (1979) also found a positive relationship between the amount of attack shown towards a simulated predator and clutch size in blackbirds. Carlisle (1985) found that the cichlid *Acquidens coeruleopunctatus* would more readily flee (from the experimenter) the fewer the number of fry there were in the brood.

Andersson *et al.*'s model conforms to the intuitive prediction that a higher degree of predation pressure should, in general, produce a higher level of parental defence. It can be used to show that this prediction holds in a variety of circumstances, for example irrespective of whether the offspring depend on their parents merely for defence from predators (e.g. in many fish species) or whether they are entirely dependent on them for survival in other ways such as feeding and incubation (e.g. in birds and mammals).

The model was also used by Andersson *et al.* to explain interspecific differences between precocial and altricial birds. They predicted that the optimal level of defence will be lower in species producing precocial young, for two reasons: the first is because the young are usually scattered in the habitat, and hence a predator will be likely to threaten only one chick at a time; the second is because the chicks will be able to secure their own protection (by camouflage) as soon as they hatch. Andersson *et al.* reviewed several studies which do, indeed, indicate a decrease in parental aggression after hatching in precocial birds but not in the few altricial species which have been studied. More evidence is needed to test this prediction fully.

The approach adopted by Andersson *et al.*, which is to view the evolution of parental care in terms of a lifetime cost–benefit analysis, has some similarities to Wells' (1981) analysis of parental care in anurans. Wells emphasised the rarity of parental care in anurans and related it to the following consideration: although parental care will increase the numbers of young that survive, a lower rate of survival would still occur in the absence of parental care. The model of Andersson *et al.* predicts that the selective advantage of parental defence will be relatively low if the probability of the young surviving in its absence is relatively high: according to Wells, this is the case for anurans, but not for most birds and mammals.

The variables incorporated into Andersson *et al.*'s model are predation pressure, numbers of offspring, life expectancy of the parents and young, and the defencelessness of the young. The principal cost of parental defence they considered was the risk of injury. There are, however, other costs such as the necessity of laying the eggs in one place (hence attracting predators), and forfeiting the production of further offspring by defending current ones (Tallamy, 1984). For example, in Tallamy's own studies of the lace bug, he found that females which consistently abandoned their eggs could lay twice as many as those which remained to guard the first batch. In many cases, however, producing further offspring entails high costs which make guarding the existing ones a more beneficial option, especially if these would be in danger in the absence of protection.

Andersson *et al.* acknowledged the importance of replacement costs in their paper, but they did not include it in their model. Replacement costs would include the cost of obtaining another mate (Maynard Smith, 1977) and would be higher in value if the number and quality of the offspring were high (Carlisle, 1985). These considerations provide the central feature of another line of functional arguments (Trivers, 1972; Dawkins & Carlisle, 1976; Maynard Smith, 1977), discussed in section 5.5. Here it is sufficient to note that Andersson *et al.*'s model would be improved if it included a variable to denote the cost of replacing the offspring instead of simply the numbers of offspring; if this cost is high (as it would be in birds and mammals as the offspring became older), defence would be proportionately greater. In general, we should expect animals with larger numbers of offspring, those which have difficulties in finding a mate, and those which engage in long and costly parental care (i.e. producing 'high-quality offspring': Carlisle, 1985), to show higher levels of offspring defence, since they would incur greater costs in replacing them (Dawkins & Carlisle, 1976). This hypothesis predicts that the most furious and persistent examples of parental aggression should occur in birds and mammals, where replacement costs are high. Isolated dramatic examples which support the prediction include the following: a moose cow attacking a bear (Svare, 1981), Eastern kingbirds travelling as far as 300 metres from the nest to attack hawks or crows (Blancher & Robertson, 1982), and birds such as black-headed gulls, arctic terns and fieldfares attacking humans (Kruuk, 1964; Lemetyinen, 1972; Andersson *et al.*1980).

Such damaging attacks will, in many species, provide a contrast with the usual form of intraspecific aggression which will not be so damaging. Where there has been selection to lessen the damage incurred in intraspecific fights (see Chapter 3), we should not expect this to apply to such an extent to parental aggression. Although there may not always be as pronounced a separation in the form of aggression used in the two circumstances as occurs between antipredator and intraspecific aggression (cf. Svare, 1981), we should expect *some* degree of separation. This does appear to be the case, at least in the few available examples: the female wildebeest uses its horns to defend its calves against predators (Kruuk, 1972), and female mice have been reported to kill male intruders which were unable to flee from the nest (Svare, 1981).

5.4 Parental aggression in relation to infanticide

So far we have considered parental aggression as specifically related to the danger from predation. However, an additional threat is posed by infanticide, which is widespread in the animal kingdom (e.g. Ridley, 1978; Hrdy, 1979; Sherman, 1981; Labov *et al.*, 1985). In reviewing studies of infanticide, Sherman (1981) suggested that it serves a variety of functions. Some of these entail the parents' benefiting from killing their own young (see also Thresher, 1985), and are therefore not relevant to the present discussion. The others, where the parents are not the killers and do not benefit, are the ones which should produce strategies to oppose infanticide.

The first function of infanticide Sherman suggested was intermale competition, which is found in animals such as lions (Bertram, 1975), langurs (Hrdy, 1979) and in rodents such as the collared lemming (Mallory & Brooks, 1978). In the polyandrous northern jacana, infanticide may result from interfemale competition (Stephens, 1982).

A second function for infanticide involves cases where unrelated young are likely to wander into the brood or litter. A third is predation (i.e. cannibalism), which occurs very widely (Ridley, 1978; Sherman, 1981; Townsend *et al.* 1984). A fourth category is the removal of current or future competitors for nest or breeding sites (Sherman, 1981; see also McLean, 1983). Finally, Sherman suggested that infanticide may have arisen as a by-product of the benefit conferred by an associated behavioural trait.

Whatever the function of infanticide, we should expect it to have led to selection pressure on the sex (or sexes) performing parental care to evolve parental aggression in the same way that predation pressure would have done. But, in this case, the principal target of parental aggression would be other conspecifics.

5.5 Sex differences

Parental aggression may be shown by one or both sexes. In some animals, notably the majority of insects and fish, parental care consists mainly of guarding the eggs or young, and is so straightforward that it can be undertaken by one parent only (Huntingford, 1984*a*; Tallamy, 1984). In other cases, such as birds, the eggs require incubation, so that two parents usually co-operate in rearing and defending the eggs or young. In order to understand why different patterns of parental aggression have evolved, it is necessary to explain sex differences in parental care generally.

Most of the available models address the question of what determines whether the male or the female is involved in parental care when it is undertaken by one sex only. In the course of presenting a broad theory which links sexual dimorphism and parental care (see Chapter 8), Trivers (1972) advanced two specific hypotheses concerning which sex shows parental care. One hypothesis involved gametic investment and the other the reliability of paternity. In the first, he argued that sex differences in parental care ultimately depend on the initial imbalance in contributions to the zygote. The female provides a greater contribution (in terms of stored food) than the male does (through spermatozoa). Trivers argued that this should lead to the male being more likely to leave the female to care for the eggs, unless there were additional factors operating to alter this imbalance. However, Dawkins & Carlisle (1976) pointed out that, logically, decisions about future parental care should not depend on previous contributions, but on which strategy would produce the highest future increase in fitness. On this argument, both sexes would be equally likely to leave the other to look after the eggs or young. Whether there would be selection to stay and provide parental care, or to desert, would depend on the cost of raising future offspring to the same age, compared with the cost of finishing rearing the present ones (and also the cost of finding another mate: Maynard Smith, 1977).

Dawkins & Carlisle argued that whether fertilisation is internal or external and the *order* of gamete release are the important variables determining which sex is likely to be left to care for the offspring. They argued that where fertilisation is internal, for example in mammals, it would be more advantageous to the male to leave. Where it is external, and the eggs are released before the sperm (which occurs commonly in water, since sperm diffuse more quickly), the female will be more likely to leave. This explanation is consistent with the more frequent occurrence of paternal care in fish, and maternal or biparental care in birds and mammals. It can also account for variations between species of salamanders (Ridley, 1978), but it does not fit a more detailed comparative analysis of non-mammalian taxonomic groups: Ridley showed that there are a number of families where external fertilisation is accompanied by *maternal* care, and others where internal fertilisation is accompanied by *paternal* care. In a detailed examination of the evidence for fish and amphibia, Gross & Shine (1981) also rejected Dawkins & Carlisle's hypothesis, on the grounds that male care remains correlated with external fertilisation independently of the opportunity for the male or female to desert the zygotes first.

Trivers' (1972) second hypothesis concerned the reliability of paternity. He suggested that paternity would be relatively certain in externally fertilised animals (especially if sperm release occurs after egg deposition), but it would be relatively uncertain in internally fertilised species. Ridley (1978) argued that this hypothesis predicts that maternal and paternal care would be equally likely in externally fertilising animals; however, he found a preponderance of paternal care instead (see also Werren, Gross & Shine, 1980; Gross & Shine, 1981). Gross & Shine (1981) argued that the data for fish and amphibians supports the 'association hypothesis', of Williams (1975), which states that prior association with the embryos preadapts that sex for parental care. With internal fertilisation, it is usually the female, whereas external fertilisation, particularly when it occurs in the male's territory, preadapts the male for care. However, this hypothesis does not explain why one or other sex has a prior association with the embryos. Ridley (1978) did pose this question in relation to why the male rather than the female should be territorial in fish. Following Trivers (1972), he suggested that it was the result of female choice of oviposition sites. Territorial males will have demonstrated their ability to defend a site already and are therefore likely to be able to defend the eggs in future. Ridley also noted that where the male has to drive off the female to prevent her eating the eggs, this also tests his ability to defend them!

The models considered so far are 'economic' ones, concerned with establishing the optimal level of lifetime fitness. In Chapter 1, we noted that the optimal strategy in social contexts will depend on the available alternatives and their frequencies, and that the concept of evolutionarily stable strategy (ESS), derived from game theory, is more appropriate here. In three simple models, Maynard Smith (1977) represented the factors which would determine which sex was more likely to desert the eggs or young in game theory terms. The three models differ in whether discrete breeding seasons or continuous breeding is assumed and whether or not a substantial contribution is made to the eggs before copulation.

In all three models, both male desertion (referred to as the duck ESS) and female desertion (the stickleback ESS) are possible, but only if one parent is effective in guarding the young and reproductive prospects are good for the deserting parent. Where there is a substantial contribution to egg production before copulation, as occurs in many fish, it tends to be the male which remains if the female can lay more eggs overall by not guarding those she has just laid. Where reproductive success varies only with the parental contribution after copulation (as in birds), and thus the female has to wait longer before being able to re-mate, it will be the female which remains. However, the models do not readily predict which of the two strategies is likely to evolve unless the initial conditions are clearly specified, and consequently they have less precise predictive value than the economic models.

Although the models considered in this section do not specifically address the issue of parental aggression, this aspect forms such an important part of parental care – the major part in many arthropods and fish – that the models can be used to address the issue of which sex shows parental aggression in particular taxonomic groups. Application of the models to more specific questions such as sex differences in the timing, strength and form of fighting shown in defence of offspring (e.g. Curio, Regelmann & Zimmermann, 1984, 1985; Lea, Vowles & Dick, 1986) is more problematic. At present, this is restricted to some general predictions, for example that females will tend to invest more than males (Trivers, 1972), and hence would show more intense and risky forms of aggression (Itzkowitz, 1985, for a cichlid). Of course, this prediction will not apply to monomorphic birds (Trivers, 1972), and Regelmann & Curio (1986) noted that among these, the male often shows more boldness in parental aggression than the female. They also noted that the model of Andersson *et al.* could not explain these sex differences, and tested three more specific hypotheses, using data from their studies of brood defence in the great tit. Here, the sex difference involved greater initial male defence, but this diminished between broods. These findings were consistent with the view that male antipredator aggression has an additional function to that of brood defence, possibly to defend the female. It was not consistent with two other models, based on sex-specific parental care and sex-specific mortality, respectively. Curio & Regelmann's research illustrates the usefulness of developing specific models in conjunction with empirical studies.

5.6 Life cycle changes

Andersson *et al.*'s model, described earlier in this chapter, contained a prediction originally made by Williams (1966) that the optimal level of parental aggression would increase as parents became older. This would apply to comparisons both from one breeding season to the next and within the same season (such as earlier and later phases of incubation). In Andersson *et al.*'s model, the prediction is based on the ratio of parental to offspring survival probability, which decreases with time. Parental aggression therefore increases with age, regardless of factors such as the offspring's likelihood of surviving on their own (i.e. whether parental care has merely a protective function or whether it is more complex). The prediction is therefore intended to be widely applicable to different taxonomic groups, from invertebrates to mammals.

The same prediction is made by Tallamy (1984) in relation to the concept of 'reproductive value' (Fisher, 1930) which is an age-specific measure of the animals' potential contribution to future generations; this decreases from the age when the first offspring are produced until it is zero at the end of reproductive life. Tallamy predicted that the intensity of parental aggression should be inversely related to the animal's reproductive value, i.e. it would increase with age.

An alternative (or complementary) prediction, that parental aggression will increase as the breeding cycle progresses, can be derived from the models of Trivers (1972), Dawkins & Carlisle (1976) and Maynard Smith (1977). It concerns the cost of replacing the offspring to their present stage of development. This will, of course, also increase as the offspring develop and will be higher in animals whose parental investment is high, such as birds and mammals. Although Andersson *et al.* acknowledged this in their paper, it is additional to the age-related factor incorporated in their model. According to them, the main difference between the two sets of predictions is that theirs would entail increased parental aggression throughout the breeding season even if it were impossible to produce a replacement brood. Such a condition could, however, be represented in the other models by a very high replacement cost, and this would lead to the same predictions.

Empirical studies generally support the prediction of increased aggression with parental age. However, most of those carried out on vertebrates have involved a single breeding season, so that it is impossible to separate parental age and replacement costs. For example, Peeke & Peeke (1982) found that the territorial aggression of convict cichlids increased as their eggs and young developed from one stage to the next. Similar changes have been found in a variety of bird species (Wilson, 1975, Andersson *et al.*, 1980, for reviews; see also Blancher & Robertson, 1980; Curio *et al.*, 1984, 1985). Lemmetyinen (1971), for example, found that both arctic terns and common terns showed increased frequency of attacks on a dummy predator throughout the incubation period, and that attacks became most violent in the fledgling phase.

'Maternal aggression' in laboratory rodents increases during pregnancy, remains high during the early lactation period, and then declines as lactation advances (Svare, 1981). This pattern clearly does not conform to that predicted by the models referred to above; it is likely that factors such as a reduced tendency of adult conspecifics to attack older young or the increased ability of the young to defend themselves would operate in this case: for example, Wolff (1985) found that little attack by *Peromyseus* conspecifics was directed to 17- to 18-day-old pups compared to younger ages; in the minority of cases where attack was observed, pups of the older age showed a defensive posture and would bite back at the attacker, which then retreated.

There is one study which did separate changes in parental aggression within and between breeding seasons. Tallamy (1982), studying the lace bug *Garagaphia solani*, found an increase in parental aggression as the mother became older (between broods), as well as an increase as the nymphs reached maturity (within broods), thus supporting the general prediction that parental aggression increases with age.

The functional models described in this section predict that there will be a general increase in parental aggression with age, and in addition that it will increase as a particular brood becomes older. At present, there are more data relevant to the second prediction than the first, the available evidence being consistent with both predictions.

5.7 Conclusions

As expected, empirical studies showed that parental aggression is associated with increased survival chances for the eggs and young. The model of Andersson *et al.* predicts higher levels of parental aggression with higher predation pressure, increased numbers of young, and in species where the young have poor survival chances on their own. Variables the model did not take into account included costs other than the risk of injury, and replacement cost, which was seen as being positively related to the level of parental aggression. It was also suggested that parental aggression, like protective aggression, would take more damaging forms than competitive (intraspecific) aggression. Infanticide is a widespread danger in the animal kingdom, and would exert a similar selection pressure for the evolution of parental aggression as does predation.

Sex differences in parental aggression were discussed in terms of general models of the evolution of sex differences in parental care. These models seek to predict the circumstances in which the male or the female will undertake parental care when only one parent is involved. The crucial explanatory variables which were used included the unequal contribution of the male and female to the zygote, whether fertilisation is internal or external, and the reliability of paternity. However, none of these could accurately explain the pattern of sex differences in a wide variety of taxonomic groups. The association hypothesis, which related parental care to prior association with the embryos most accurately fitted the evidence, but did not explain why one or other sex would show more prior association. Game theory models of whether biparental, male or female care would be an ESS included variables such as reduced mortality resulting from single-sex rearing, and the chances of finding another mate on deserting. These were both found to influence whether there is biparental or single-sex care in animals where pre-copulatory parental investment is small (such as birds). Where it is larger, as in many fish, male care is more likely because the female can increase its fitness by moving on to lay more eggs, rather than guarding.

In section 5.6, Andersson *et al.*'s model was used again, to predict that parental aggression would increase as the parents became older, but the same prediction could be derived from Fisher's concept of reproductive value. An additional or complementary prediction was that parental aggression would increase throughout a single breeding season owing to the increased replacement costs involved. Most empirical evidence is confined to a single breeding season (and generally supports the prediction).

5.8 Suggestions for future research

At present, there are a number of studies which assess survival rates in the presence or absence of parental aggression (section 5.2), but relatively few which test directly the predictions of the functional models described in this chapter. More integration between the theoretical models and empirical research would be a useful general guide for future work. Questions posed by the theoretical models can be answered in several ways, from field studies by comparative evidence, and by correlational data on survival rates, and from experimental studies measuring survival rates of offspring.

Specific predictions from Andersson *et al.*'s model which require further testing include the difference between precocial and altricial species, and further assessment of the relationship between brood size and parental aggression (section 5.3). The model itself also requires further refinement, for example to include replacement costs as well as numbers of offspring, and to consider sex differences in the pattern of defensive aggression within a brood (Lea *et al.*, 1986) and between broods (Curio *et al.*, 1984, 1985), although the last of these may require alternative models (Reglemann & Curio, 1986). The prediction that parental aggression increases with age across breeding seasons can be investigated using the method adopted by Tallamy (1982) in his study of the lace bug.

One topic which has received very little attention is the suggested separation of competitive and parental aggression in species where damage is likely to be inflicted (analogous to that shown for protective and competitive aggression). A detailed investigation of target sites and weapon use, by the same animals in the two circumstances, would provide evidence on which to assess this hypothesis.

The models put forward to predict the circumstances in which one or other sex will guard the eggs and young are more difficult to test. At present, discussion of these theoretical models has been restricted to assessing them on the basis of general comparative evidence taken from a wide range of species. Eventually, such general surveys should give way to measuring correlations between features on the model or models and specific behavioural outcomes.

In relation to parental aggression, fitness consequences are relatively straightforward to assess (compared with the investigation of feeding behaviour or exploration, for instance), since offspring survival is involved. However, measuring offspring survival in the face of predation or infanticide poses ethical problems (Huntingford, 1984*b*). If assessments are to be made of the fitness consequences of different patterns of parental aggression, observations of naturally occurring variations are ethically preferable to experimental studies; and experiments on survival consequences carried out on species which defend eggs, or on invertebrates, are preferable to those using vertebrates with well-developed young (cf. Huntingford, 1984*b*). Invertebrates have the additional advantage of producing more offspring in relatively shorter periods of time than vertebrates.

One way of investigating parental aggression in vertebrates which overcomes some of the ethical objections to presenting them with live predators is to use a model. Another is to use a human intruder. Some of the studies referred to in this chapter have adopted one or other of these procedures (e.g. Lemmetyinen, 1971; Robertson & Bierman, 1971; Carlisle, 1985). In order to assess the significance of these alternative 'predators' in relation to parental responsiveness, Knight & Temple (1986) examined whether the nest defence of red-winged blackbirds differed in response to three sorts of 'predator': a live or mounted crow, different postures shown by a human intruder, and unfamiliar and familiar human intruders. They found that a higher intensity of mobbing attacks was directed towards the mounted crow (which did not move or look at the blackbirds), than to the live crow. They also found that higher intensities of attack were directed towards a human intruder who stared at the birds than to one who did not, and towards a familiar rather than an unfamiliar human intruder. Knight & Temple suggest that the implications of their results for assessing previous studies are that those using live and mounted predators may not be comparable, and the eyes of the predator may be important (in their study inhibiting attack but, in others, eliciting mobbing: e.g. Curio, 1975). More generally, it is suggested that reports of existing avian studies have not controlled for a number of variables such as the species of predator, age, breeding state, previous experience, territory, weather and habituation. Variables found to be important in Knight & Taylor's study, such as the posture of the predator, were incompletely described in about half of a sample of recent papers they reviewed, and many studies have relied on only one type of antipredator response such as an alarm call.

The potential benefits of future research taking into account these methodological inconsistencies would be very great in terms of providing a much more solid and consistent basis for an overall comparative assessment of avian parental aggression than is possible from present studies. Nevertheless, it is easy to be critical of the methodology involved in field studies, where practical considerations often dictate what can realistically be achieved. It is doubtful whether it would be possible to obtain a consensus from all potential field workers concerning the methodological details and the variables which should be controlled in every study. We are at the stage where there are relatively few studies on which to make comparative assessments, and for many species there are *no* existing studies. Of course, methodologically sound and consistent studies would be preferable to those which are inconsistent or flawed, but even these would be better than no studies at all.

Overall, future research on functional aspects of parental aggression should involve an integration of the theoretical models with comparative and correlational field studies, which take account of the methodological issues raised above, and with experimental studies which take into account the need to avoid or minimise suffering to the young.

Further reading

Andersson, M., Wicklund, C. G. & Rundgren, H. (1980). Parental defence of offspring; a model and an example. *Animal Behaviour*, **28**, 536–42.

Ridley, M. (1978). Paternal care. *Animal Behaviour*, **26**, 904–32.

Svare, B. B. (1981). Maternal aggression in mammals. In *Parental Care in Mammals*, ed. D. J. Gubernick & P. H. Klopfer, pp. 179–210. New York: Plenum.

Tallamy, D. W. (1984). Insect parental care. *Bioscience*, **34**, 20–4.

6

Mechanisms of parental aggression and their relation to function

In this chapter, we consider the mechanisms through which parental aggression achieves its function. First, we discuss the general design features underlying parental aggression in various taxonomic groups. This is followed in section 6.2 by an outline of the general principles underlying maternal aggression in rodents, the form of parental aggression which has been most extensively studied in terms of mechanisms. Internal factors controlling the beginning and end of parental aggressiveness are discussed in section 6.3. In the following three sections, interspecific and interindividual differences are considered in terms of mechanisms responsive to adaptive change at the phylogenetic, developmental and short-term levels, following the distinction made in Chapter 1.

6.1 General principles

The function of parental aggression is to protect the eggs or young from predators or conspecifics. We should therefore expect similar mechanisms to those involved in protective aggression, with the possible addition of ways of assessing the costs and benefits of various behavioural options in the short term. We should also expect mechanisms for restricting parental aggression to the appropriate phase of the life cycle or alternatively to when eggs or young are present in the immediate environment.

In the case of parental aggression, there are no examples of the very simple mechanisms outlined in Chapter 4 for protective aggression (found in sea anemones and other hydrozoan coelenterates). Defence of the eggs and young is usually restricted to those invertebrate groups showing more pronounced encephalisation (Ridley, 1978; Tallamy, 1984).

In vertebrates such as fish and birds, as well as in some insects, parental aggression may overlap with territorial behaviour. As we noted in the previous chapter, in many teleosts, defence by the male is a common consequence of prior territorial behaviour associated with nest building. In birds, territorial behaviour is typically shown by the male prior to parental behaviour, which usually involves both sexes. Nevertheless, territorial and parental aggression may be similar in that they are both directed towards intruders into a spatial area: however, they are directed towards different types of intruder, and they may involve different types of response (see discussion at the end of section 5.3).

The motivational separation of general antipredator responses and parental antipredator attack in birds can be illustrated by considering Curio's study of the pied flycatcher, described in Chapter 4. He found that snarling attack was specifically shown to predators which preyed only upon the nestlings whereas more dangerous predators were mobbed. Curio argued that a series of matching mechanisms (IRMs, see Chapter 4) operates in parallel in the detection of nest predators. Each one is set to respond to a specific type of stimulus configuration, but all of them produce the same range of responses. We noted earlier that a novelty-detecting (mismatch) mechanism was also involved.

Although much less common than in birds or fish, there are some examples of mammalian parental aggression taking the form of male territoriality. McLean (1983) found that male arctic ground squirrels defended territories at the time when young which they were likely to have sired were at risk from strange males.

In many mammalian species, the female but not the male shows parental care. Where the main danger comes from conspecific males, there will be a complete separation of territorial and parental aggression. Only the female will show parental aggression in such circumstances, and its onset and ending are controlled by factors totally separate from male territorial aggression.

Most of the experimental studies of parental aggression have been restricted to rodent species which show these features, such as the housemouse, the hamster, the rat (Svare, 1981), the collared lemming (Mallory & Brooks, 1978) and three species of *Peromyscus* (Eisenberg, 1962; Gleason, Michael & Christian, 1980; Wolff, 1985). In most cases, aggression occurs during pregnancy and during the lactation period (Ostermeyer, 1983).

6.2 General principles underlying rodent maternal aggression

The form of attack shown by a female rat or mouse to an intruding male appears similar to that shown by an adult male attacking a strange conspecific which has entered its familiar area (Svare, 1981). We should, however, note reports of severe and damaging attacks by a female with a litter (Svare, 1981), and the more general evidence that this will be a more damaging form of attack than competitive aggression (section 5.3).

The maternal aggression of rodents shares the following features with territorial aggression: attacks are directed to a wide range of novel intruders, a decline in attack rate is shown with repeated exposure, and aggressive responses are dependent on the presence of a familiar area (e.g. Mallory & Brooks, 1978; Gleason *et al.*, 1980; Svare *et al.*, 1981; Wolff, 1985). Territorial attack can be represented in general terms by the discrepancy model outlined in Chapter 4. Fig. 6.1 shows how this general model can be applied to parental aggression: in this case, the input is monitored for mismatch from a stored representation of a spatial area containing the eggs or young (or simply a spatial area in the case of the prepartum aggression of mammals). Decision process 1 is influenced by a variety of factors which affect aggression generally (e.g. the strength of the input and past experience), and also by those which are specific to this type of aggression, such as hormones indicative of reproductive state and an input indicating the presence of eggs or young. These influences will be discussed in more detail later in this chapter.

In Chapter 4, we elaborated the suggestion that variations in the degree to which a given input signal produces attack rather than fear could underlie interspecific and intersex variations in protective aggression. This variable ('boldness', from Huntingford, 1976a) would need to be generally high in the case of parental aggression. Indeed, the decision process 1 in Fig. 6.1 would be heavily biased towards attack, provided the specific features controlling the onset of parental aggression were present. Andersson *et al.*'s model predicted that predation pressure (including danger from conspecifics), the number of offspring, and the defencelessness of the young would all increase the likelihood of attack towards an intruder (as would the cost of replacement: Dawkins & Carlisle, 1976). These functionally relevant variables would all contribute towards increasing the level of input at which flight replaced attack for particular species. In some animals, this will take the form of a flexible mechanism enabling the decision to be

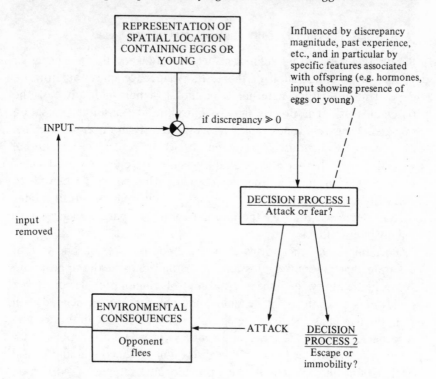

Fig. 6.1. Control system model of aggression (Fig. 4.2) depicting parental aggression.

based on an assessment of the degree to which such variables are present at the time (see section 6.6).

The mismatch mechanism shown in Fig. 6.1 is based on studies showing that attack is dependent on the presence of a familiar area, that novel intruders are attacked and that attack declines with repeated exposures (see above). However, there are some data which indicate specific biases to attack certain types of stimuli more readily. In mice, males are generally attacked more than females (Svare & Gandelman, 1973; Ostermeyer, 1983; Elwood, 1985) and young without hair are rarely attacked (Svare & Gandelman, 1973): this reflects the relative danger to the female and its offspring, since unmated males tend to be infanticidal and dominant males are more infanticidal than subordinates. These findings indicate that a more complex comparison process – perhaps involving an assessment of the perceived degree of threat posed by the intruder – is involved in addition to the mismatch mechanism shown in Fig. 6.1.

6.3 The internal control of parental aggressiveness

Parental aggression, by definition, occurs only when there are eggs or young, or in anticipation of these. Its occurrence must therefore be dependent on either external or internal signals indicative of the parental phase. The precise signals controlling the onset and cessation of parental aggressiveness appear to vary greatly in different animals. One major distinction is the extent to which control is by internal variables, or whether the animal responds to environmental cues. Control by environmental cues – particularly where these can be used to assess the costs and benefits of fighting – will be discussed in a later section. Here we consider the internal control of parental aggression, mainly by hormones.

Hormonal and other forms of internal control provide one way of ensuring that parental aggression begins and ends at the appropriate stages in the life cycle. Since internal controls cannot provide very direct indications of changes in the immediate environment associated with reproductive phase, such as the age and numbers of young, they can be regarded as one of the more indirect ways in which a functional end-point is achieved. In terms of the discussion of functional and causal explanations in Chapter 1, they provide an example of correlated characteristics, where the principles underlying the mechanism are unrelated to the adaptive consequences.

There are few studies providing data on the internal mechanisms controlling parental aggression in invertebrates. One exception is the work of Loher & Huber (1966) showing that the female grasshopper *Gomphocerus rufus* kicks away approaching animals within 2 hours of copulation. This change in behaviour is signalled by feedback from the spermathecal duct (the seminal receptacle), reflecting the volume of spermatophore inserted by the male.

Virtually nothing is known about the possible role of hormones in the control of parental aggression in non-mammalian vertebrates (cf. Harding, 1983; Villars, 1983). There is, however, one study carried out on the ring dove, where hormonal control of reproductive and parental behaviour has been well studied. In the female, gonadotrophin secretion stimulates oestrogen and progesterone secretion, which leads to nest building and the onset of incubation. The latter produces, in turn, prolactin secretion which maintains incubation and stimulates crop milk production (Slater, 1978; see also Lea *et al.*, 1986). Vowles & Harwood (1966) administered either prolactin or progesterone (together with oestrogen or alone) and found that both hormones would increase the defensive aggression shown by female (and male) doves to a potential predator. In females, these hormones also increased defensive aggression towards other doves. From these results, it appeared likely that parental aggression in the ring dove is to some extent directly controlled by the internal secretion of reproductive hormones. More recent studies (e.g. Lea *et al.*, 1986) have found sex differences in the expression of nest defence, with the female showing higher levels at the time of laying, which coincides with a marked increase in progesterone, and the male showing only a low level at laying, but a steady increase up to the time of hatching. Lea *et al.* suggest that both sexes may then be responding to prolactin.

Earlier studies of the laboratory housemouse produced some inconsistencies in whether or not it showed prepartum aggression (Goyens & Noirot, 1977; Svare, 1981). More recent evidence presented by Ostermeyer (1983) indicates that there are two forms of maternal aggression, the first represented by a steep increase in the number of attacks during the first 5 days of pregnancy, which falls to a lower level later (Figs 6.2 and 6.3), and the second by a renewed increase after parturition.

The first of these is comparable to increases in aggression following mating which have been observed in the parental sex of other animals,

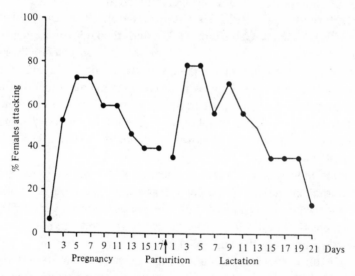

Fig. 6.2. Percentage of pregnant and lactating female housemice that attacked unfamiliar male conspecifics, at different stages of pregnancy and post-parturition. (From Ostermeyer (1983), p. 157.)

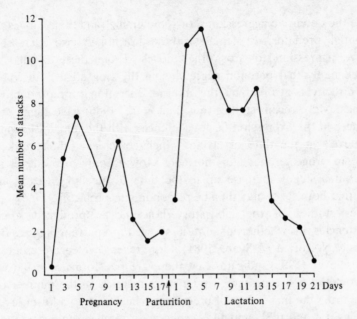

Fig. 6.3. Mean number of attacks by pregnant and lactating female housemice upon unfamiliar male conspecifics, at different stages of pregnancy and post-parturition. (From Ostermeyer (1983), p. 158.)

for example the species of grasshopper studied by Loher & Huber (1966), the three-spined stickleback (Sevenster-Bol, 1962), the golden hamster (Lisk, Ciaccio & Catanzaro, 1983), and the bank vole(Alder *et al.*, 1981). Aggression at this time is likely to function principally in providing an area free from intruders in anticipation of egg laying, egg development or parturition. However, Elwood (1985) has suggested a more specific function for the post-copulatory rise in aggression by the female housemouse, which is to subordinate the father, thus inhibiting the general infanticidal tendency shown by unmated males, replacing it with parental care.

In the housemouse, and in other species where the pregnant female shows maternal aggression (e.g. golden hamster, deermouse, vole), there are few studies on its possible hormonal basis (Ostermeyer, 1983). Noirot, Goyens & Buhot (1975) have suggested that ovarian hormones are involved, principally the high level of progesterone relative to that of oestrogen during early pregnancy. This suggestion was tested by Svare, Miele & Kinsley (1986), who hysterectomised mice during pregnancy, thereby removing their source of progesterone. Such mice showed very little aggression to male intruders (whereas sham-operated animals retained their high levels of attack); progesterone implants restored the aggressive behaviour of hysterectomised mice and simultaneous exposure to oestrogen had an inhibiting effect. These results, therefore, support the involvement of progesterone in the prepartum aggression of the housemouse.

Reviewing studies of possible hormonal influences on the postpartum aggression of the housemouse, Svare & Mann (1983) concluded that 'unlike other aspects of agonistic behaviour, like intermale aggression, the role of hormones appears to be more indirect and less dramatic'. There appears to be an interaction between internal hormonal changes and external stimulation, the latter ensuring that there is some degree of adaptive flexibility in the behaviour.

Parental aggression occurring during the postpartum period is, according to Svare (1981), at its most intense during days 3–8, and declines as the lactation period advances (see Figs 6.2 and 6.3). Comparable results have been obtained for rats where the highest frequency of attacks and bites occurs on postpartum day 9 (Svare, 1981). In several other species, maternal aggression declines during the second half of the lactation period (Ostermeyer, 1983).

During the lactation period, maternal aggression depends upon the presence of the pups. For rats, mice and hamsters, removal of the litter significantly decreases fighting (Svare, 1981; Giordano, Siegel & Rosenblatt, 1984). In mice and hamsters, it can be restored by replacing the pups for a short period, and direct physical contact is not necessary for this to occur in mice.

In a series of experimental studies of mice (e.g. Svare & Gandelman, 1973), it was shown that the crucial factor for the occurrence of maternal aggression is the experience of nursing the young (but not necessarily involving lactation), and that the minimum period of nursing for producing normal levels of fighting was between 24 and 48 hours (Svare, 1981).

Virgin female mice can be induced to nurse young by being given oestrogen and progesterone daily for 19 days, and they too show 'postpartum' aggression in these circumstances (Svare, 1981). They also stop fighting when the pups are removed and resume fighting when they are replaced.

Svare (1981) proposed that there are three phases to the control of postpartum aggression in the mouse. During gestation, oestrogen and progesterone produce nipple enlargement, which prepares the animal for suckling stimulation during the postpartum period, and a minimum period of this stimulation initiates aggression; finally, its intensity declines as lactation progresses. The role of hormones would therefore appear to be restricted to a preparatory one in these studies of postpartum aggression.

In terms of the motivational model outlined in the previous section (Fig. 6.1), the influence of hormones during pregnancy will be a relatively direct one on decision process 1, to increase the likelihood of attacking an intruder in the area surrounding the nest. This could be achieved either by an increase in 'boldness', as suggested earlier, or else by an increase in general reactivity. In the second case, the input signal (degree of discrepancy in Fig. 6.1) necessary to activate decision process 1 would be reduced (as shown in Fig. 4.7). In this way stimuli which would ordinarily produce only an orienting response and investigation, would, during pregnancy and lactation, evoke attack. It seems likely that both these changes would be necessary, since the second by itself would also produce an animal which more readily shows fear at higher input levels.

During the postpartum period, similar changes are brought about through a mechanism which is also responsive to the appropriateness of external circumstances in the general sense that it is dependent on the presence of the young.

There is, as indicated above, an indirect relationship between the hormonal changes producing prepartum aggression and its function (to provide an undisturbed area in anticipation of the arrival of the pups). During the postpartum period, the relationship between function and mechanism becomes more direct: the functional consequence of postpartum aggression is to provide an undisturbed area resulting in an increase in the pup's chances of survival: in this case, a mechanism which is sensitive to whether or not they are still present incorporates at least the major functionally relevant variable into its control system. In section 6.6, we consider other cases – mainly in non-mammalian vertebrates – where there is evidence of more extensive environmental control of parental aggression.

6.4 Mechanisms underlying individual differences in responsiveness

We now consider mechanisms which could account for some of the differences between species and the sexes discussed in Chapter 5. Again, much of the experimental evidence is derived from the maternal aggression of rodents: interspecific and interstrain differences are found in the typical level of maternal aggression, its latency and in its timing.

Some strains of mice show high levels of maternal aggression whereas others show little or none. Ebert (1983) has shown that female housemice can be selectively bred for high and low levels of maternal aggression (which was significantly correlated with interfemale aggression). These differences in the general level of maternal aggression could be accounted for in terms of different degrees of 'boldness' shown in response to the same level of input, as outlined in section 6.2.

In some species, such as the housemouse, attack occurs shortly after the detection of a strange conspecific (Svare, 1981), whereas in others (e.g. the Norway rat and the golden hamster) latencies to attack are longer. This time is occupied by social investigation of the intruder, which presumably involves stimulus-matching mechanisms on the input side (as was suggested, in Chapter 4, for a male colony rat investigating an intruder). One possible functional explanation for these species differences is that they are related to the degree to which all or only some classes of conspecifics pose a threat to the young: animals would presumably not spend a long time investigating a predator or a conspecific which was clearly cannibalistic, and in the case of the housemouse, unmated males generally tend to be infanticidal towards pups (see above). It is therefore possible that the comparison process preceding attack has been subject to selective modification in relation to the degree of danger to the young (one of the variables in Andersson *et al.*'s model), so that increasing likelihood of danger has resulted in a simpler set of comparison rules, and hence a more rapid – but less discriminating – response. In this case, we should also expect decision process 1 (Fig. 6.1) to be heavily biased towards attack, through the influence of hormonal and other factors which control the readiness to show parental aggression.

The timing of maternal aggression may vary in different rodents. In some species, such as the housemouse, prepartum aggression may be more pronounced early in gestation, whereas in the golden hamster and the deermouse, it appears to be more pronounced during the second half of gestation, although these tentative conclusions are based on only a few studies (Ostermeyer, 1983; see also Giordano, Siegel & Rosenblatt, 1986).

In terms of the distinction between the time-scale of adaptive mechanisms made in Chapter 1, at least some of the species and strain differences outlined in this section represent adaptations whose flexibility lies at the phylogenetic level; i.e. they rely on mechanisms which are relatively unresponsive to developmental or short-term changes. Thus, the interstrain differences in maternal aggression referred to above are found to persist despite different rearing environments (Svare & Mann, 1983). Whether other species and strain differences are of this nature must await further studies of the developmental influences on maternal aggression. Nevertheless, it is likely that the pronounced differences in parental aggression between different taxonomic groups are dependent on major differences in the neuroendocrine control of parental aggression which, as outlined in section 6.3, has been little studied outside the rodents.

6.5 Developmental flexibility

There has been very little research into developmental influences on parental aggression. One exception concerns the study of prenatal experience.

vom Saal (1984) investigated the influence of intrauterine position on aggression in the housemouse. In relation to parental aggression, it was found that lactating females which had been positioned between two males in the uterus showed more intense aggression towards a nest intruder than lactating females whose prenatal position had been between two females.. vom Saal suggested that the mechanism underlying this effect is hormonal, since female mice housed between two males had higher androgen levels in foetal life.

It is also known that prenatal stress (consisting of heat, restraint and bright lights) influences adult maternal aggression in mice, on this occasion reducing the number of attacks on a male intruder during the first lactation period (Politch & Herrenkohl, 1979). Again, the effect may be mediated via changes in foetal androgen levels, since prenatal stress has been found to produce dramatic increases in foetal testosterone levels on day 17 (vom Saal, 1984).

vom Saal has argued that these early developmental influences represent an adaptive mechanism to produce a form of phenotypic polymorphism which can adapt to different degrees of population density and competition. His adaptive argument will be discussed in Chapter 8.

There is a single study of early postnatal social influences on parental aggression. This showed that postweaning isolation did not influence the development of aggression by lactating housemice (Svare *et al.*, 1981), in contrast to the results described for shock-induced aggression in Chapter 4 (and also competitive aggression: e.g. Valzelli, 1969).

Again, only isolated examples are provided by these developmental studies, once more indicating the lack of research on the development of aggression.

6.6 Short-term mechanisms

In section 6.3, we discussed postpartum aggressiveness in rodents and showed that it was to some extent under hormonal control, yet it was also responsive to whether young were present. As indicated there, such a mechanism shows relatively little fine-tuning in its adaptive capabilities. This may not, however, be typical of parental aggression in other taxonomic groups. Discussing, in general terms, intraspecific variations in parental care, Carlisle (1982) concluded, mainly from studies of fish and birds, that there is a widespread ability to adjust the level of parental care to the prevailing conditions. Curio *et al.* (1984) also suggested that nest defence among great tits is gauged to the quality of the young.

These examples suggest that the level of parental aggressiveness is responsive to changes in the relative costs and benefits indicated by cues from the young. In some cases, other environmental cues may determine whether or not defence of the eggs or young occurs at all. Wells' (1981) review of parental care in anurans indicated some degree of responsiveness to ecological conditions in members of this group. Cases of facultative parenthood were pointed out, where the adults would remain with the young only when environmental conditions were adverse, notably when there were a large number of predators. Wells also gave an example of an unpublished study by Kluge on *Hyla rosenbergi*. In this species, when population density is high, most males guard the nest against conspecifics, whereas at low densities the males rarely guard the nest.

In the lace bug, *Gargaphia solani*, a different form of facultative parental defence has been observed. Wherever possible, females oviposit in recently established egg masses of conspecifics, which inadvertently guard the eggs of the donors, thus freeing them to lay more eggs instead of guarding (Tallamy, 1985, 1986).

In some fish species, facultative parenthood may be found in one sex only, the male. Keenleyside (1983) found that males of the monogamous cichlid *Herotilapia multispinosa* frequently deserted their mates in experimental ponds where the sex ratio was biased towards females. Townshend & Wootton (1985) also observed male desertion in a population of another monogamous cichlid, *Cichlasoma panamense*. In this case, comparison with another population where only biparental care was observed suggested that predation pressure, the value of the current brood and the sex ratio all influenced the male's decision whether or not to leave the female to guard the offspring. These influences would all be expected from the functional models discussed in Chapter 5. The important point for the present discussion is the males' behavioural flexibility, showing either parental defence or no parental care, depending on an assessment of the current environmental conditions.

Studies such as these could be extended, so as to provide more systematic and quantitative evidence of the degree to which the occurrence of parental aggression is dependent upon, or influenced by, cues in the immediate environment indicative of the relative costs and benefits of defending the eggs or young. Such studies could adopt an approach analogous to that which has been used to study territorial behaviour in nectar-feeding birds, which are known to abandon territorial behaviour when nectar levels are high and to resume it again when they become low.

The studies reviewed in this section so far indicate that in some species, the decision whether to show parental defence at all is made on the basis of current environmental circumstances. Such flexibility would be expected only in taxonomic groups where there is the possibility of eggs or young surviving in the absence of parental care, or where there was the possibility of 'deceiving' another parent into looking after them. In birds and mammals, and in many fish, we should not expect facultative parenthood since there is no possibility that eggs or young can survive without parental care and there are few opportunities for deception. Instead, flexibility takes the form of different intensities of parental aggression in accord with indicators of the numbers and quality of the young.

The experimental studies of Robertson & Biermann (1979), and of Carlisle (1985), referred to in Chapter 5, provide the most direct evidence that this can occur. Robertson & Biermann manipulated clutch size in the red-winged blackbird and measured nest defence to dummy predators. They found significantly higher levels of aggression with increasing clutch size (although no consistent relationship between brood size and nest defence aggression). Carlisle found that the cichlid *Aequidens coeruleopunctatus* showed greater readiness to abandon its brood in response to a fear-evoking stimulus when a smaller number of fry were in the brood. In both cases, direct estimates must have been made by the parent of the current size of the brood, since the changes in behaviour occurred in response to eggs or fry being removed from or added to the brood. We should, however, note that the methodological reservations associated with the use of a dummy predator or a human intruder, discussed in section 5.8, would apply to both studies.

The view that current parental behaviour is based on past investment (Trivers, 1972) was shown to be incorrect as a *functional* explanation (section 5.5). Nevertheless, since past investment and future prospects are often correlated, it does appear to operate as a *mechanism* in several species (Carlisle, 1985; Coleman, Gross & Sargent, 1986). This would provide a direct contrast with the assessments of current conditions being made in the two studies just described. Here, past parental investment was the same in all the experimental groups, and only current conditions were manipulated as the dependent variable.

In other cases, it is known from observational studies that parental defence increases as the young become older (Lemmetyinen, 1971; Andersson *et al.*, 1980; Peeke & Peeke, 1982; Tallamy, 1982). Whether in these cases there is a direct response to cues provided by the young or whether the mechanism is controlled by the level of past investment must await further studies. Carlisle (1985) suggested that the degree to which parental behaviour is sensitive to current conditions would depend on the nature of the animal's past environments: where these had been predictable, previous investment would have been a good indicator of future prospects, but where they had been unpredictable, it would be a poor predictor.

In terms of the motivational control system outlined earlier (Fig. 6.1), a change in the characteristic described previously as 'boldness' could be operating in Carlisle's study, where readiness to flee was increased or decreased as the numbers of fry increased or decreased (direct measures of attack were not made in this study). In the model shown in Fig. 6.1, assessment of relevant external cues could be shown as a factor influencing decision process 1, to influence the point at which fear overtakes attack, in a similar way to that suggested earlier for hormonal influences.

Changes in responsiveness to indicators of the immediate environmental cost–benefit requirements can therefore be represented as a form of short-term flexibility which is built into the particular motivational system. Instead of being affected by an internal signal associated with the parental phase (such as a hormone), or by some indicator of past parental investment, the system responds to an external signal providing a more direct estimate of relevant current conditions.

The discussion of short-term adaptive mechanisms in Chapter 4 outlined two others which might apply to protective aggression: first, the hormonal consequences of winning or losing a fight, which can subsequently affect behavioural responsiveness, and secondly, learning mechanisms involving classical and operant conditioning. There is no evidence for either of these in the case of parental aggression.

It is important that parental aggression results in the intruder being driven off. Whenever this does not occur, the negative consequences for fitness are great. Compared with competitive aggression, we should expect parental aggression to rely much less on mechanisms which tend to increase or decrease the likelihood of behaving aggressively according to whether the animal had won or lost a previous encounter or encounters. Generally, for competitive aggression, the fitness consequences of winning in any individual case will be of a lesser magnitude. It will therefore be more important to estimate the *cost* of fighting compared to the loss of fitness entailed by withdrawing. Conversely, the part played by mechanisms which indicate the previous consequences of fighting, assessments of the value of a resource and the fighting abilities of opponents, will be much less in the case of parental aggression. It is therefore unlikely that parental behaviour will be much affected by behavioural changes influenced by the hormonal consequences of fighting, or by the reinforcing properties of winning and the punishing properties of defeat (Chapter 8).

Nevertheless, it is possible that there may be *some* degree of modification of fighting behaviour as a result of its consequences in animals protecting their young. It may, for example, be possible to induce strains of mice low in maternal aggression (Ebert, 1983) to become more aggressive through a series of easy victories. Such studies have not yet been carried out.

Similarly, there are no studies of classical conditioning of this form of aggression. Classical conditioning would widen the range of cues to which the aggression would occur. From a functional viewpoint, this would be a more likely form of learning to operate in the case of an animal with young reacting to a predator or a dangerous conspecific. Any regularity in a cue preceding the presence of such a predator or conspecific should lead to classical conditioning of that cue. It would certainly seem adaptive for parents defending young to possess such mechanisms.

6.7 Conclusions and suggestions for future research

In this chapter, I have considered the mechanisms of parental aggression in relation to functional considerations. It is clear from the general survey in section 6.1 that parental aggression has been considered more in terms of its function than its causation, which has been relatively neglected apart from the fairly recent interest in rodent maternal aggression. Studies of mechanisms in other species of mammals, other vertebrates and in invertebrates are very sparse.

In rodents, the mechanisms underlying parental aggression were viewed as being broadly similar to those for territorial aggression, i.e. it was dependent on the attacker being in a familiar area, and novel intruders were attacked most readily (although there were also specific stimuli which were more likely to induce attack).

One important question is what controls the beginning and the end of parental aggressiveness at the appropriate stages of the animal's life cycle. Even in rodents, there is relatively little direct evidence bearing on this question. The behavioural endocrinology of prepartum aggression is only just beginning (see Svare *et al.*, 1986). The hormonal mechanisms responsible for its onset and maintenance in different species remain speculative. Several decades of research into sex hormones and behaviour in rodents has largely bypassed the study of parental aggression. Using progress in other areas as a guide, future research should aim to find out the following: first, the exact nature of the hormonal control in different species; secondly, the sites of action in the brain; and thirdly, whether they are subject to organisational influences during prenatal or early postnatal development. Existing studies of the influence of intrauterine position on the maternal aggression of mice suggest that prenatal androgenic influences extend to maternal aggression, although the precise direction of the influence is not clear (see also the study by Wagner, Kinsley & Svare, 1986, on the influence of prenatal progesterone).

There is more experimental evidence on postpartum aggression in rodents, which has been investigated since the early 1970s. Studies carried out since then show that it occurs as a result of a combination of indirect hormonal effects and cues indicating the presence of pups.

Evidence on species and strain differences was again largely restricted to rodents, but it was found that such differences occurred in the level of aggression, its latency and its timing during the reproductive phase. The extent to which such species and strain differences are modifiable through developmental influences is again uncertain, as the relevant studies have not been undertaken. Developmental studies were largely restricted to the investigation of how prenatal influences (intrauterine position and stress) influence later maternal aggression.

Short-term adaptive mechanisms were discussed in terms of facultative parenthood, which occurred in species where the eggs could develop in the absence of parental care or where the parent could deceive a conspecific into looking after them. In other cases, the *intensity* of parental aggression was influenced by factors such as the numbers and the quality of the young. In two experimental studies this was shown to be the result of current stimulation from the young, although in other cases, apparently similar observations may result from the influence of past parental investment (which will be related to current circumstances). Again, this is an area for further investigation.

The study of learning influences on maternal aggression has been completely neglected. Classical conditioning would enable animals to anticipate the presence of a frequent intruder, and therefore would be likely to operate in the case of parental aggression. It was argued that drastic modification of behaviour as a consequence of winning or losing fights would not play such an important part in parental aggression as in the case of competitive aggression.

In conclusion, causal studies of parental aggression have been very much neglected until recently, and where this subject has been studied, the investigations are largely restricted to postpartum aggression in rodents. A broad perspective covering a wider range of taxonomic groups, combined with a recognition of functionally relevant variables in such causal investigations, is recommended for the future. The ethical considerations outlined at the end of Chapter 5 will, of course, also apply to studies of the causal mechanisms underlying parental aggression.

Further reading

Ostermeyer, M. C. (1983). Maternal aggression. In *Parental Behaviour of Rodents*, ed. R. W. Elwood, pp. 151–79. New York: Wiley.

Svare, B. B. & Mann, M. A. (1983). Hormonal influences on maternal aggression. In *Hormones and Aggressive Behavior*, ed. B. B. Svare, pp. 91–104. New York: Plenum.

7

Competitive aggression: a functional analysis

7.1 General introduction to competitive aggression

In the next four chapters we consider competitive aggression. Of the three functional categories outlined earlier, most research has been carried out on this form of aggression. Indeed, the majority of topics outlined in the historical review in Chapter 1 – motivation, territorial fighting, dominance, hormonal influences, frustration, and game theory models of fighting strategies – all concern some aspect of competitive aggression.

So far, when we have considered protective and parental aggression from a functional viewpoint, we have been concerned with forms of aggression whose fitness benefits are clear and straightforward: if a predator or a conspecific is likely to kill the individual or its offspring, a high gain in fitness occurs as a result of being able to prevent this. There may be a question of assessing how much risk to undertake with particular types of predators, or with different numbers and quality of offspring, but generally such assessments are relatively limited in scope. However, when an animal competes with a conspecific over a resource such as food or a mate, the potential loss in fitness as a result of losing the encounter is usually nowhere near as great, since neither life nor young are threatened in such a fight. There is, therefore, much more scope for adopting the strategy of 'living to fight another day'. Losing one encounter over a particular mate or patch of ground or food source will usually not be as important as losing one's life or one's offspring's lives. Whether to fight or to withdraw becomes, as a consequence, much more a matter of decision-making, choice being based on an assessment of whether the benefits outweigh the costs in a given set of circumstances.

In general, there will be two sets of decisions to make in the case of competitive aggression: the first is whether it is worth fighting over the particular resource (i.e. will it contribute to fitness?); the second is whether it is worth fighting the particular opponent, and if so, what strategy to use. In this chapter and in Chapter 8, we restrict our discussion to the first of these questions, 'When do animals fight over resources?' The second question is much more complex, since it concerns assessing the ability and potential of the opponent. Its functional analysis concerns the application of game theory models to fighting strategies, a subject which has attracted considerable interest since the original paper on the logic of animal contests by Maynard Smith & Price (1973), referred to in Chapter 1. Game theory analysis of animal conflicts will be treated separately, in Chapters 9 and 10.

The enormity and diversity of the scientific literature on competitive aggression presents a problem in covering and organising the material. In part, this task is simplified by considering game theory approaches separately, as indicated above. In this chapter, and the next one, I shall retain the broad framework used in previous chapters. First, we consider briefly the concept of competition, and then outline the various forms of competitive aggression in the animal kingdom. Thirdly, we discuss the functional rules governing interspecific diversity in competition for food resources. We then consider functional analyses of the range of sex differences in aggression, and discuss life cycle changes. Chapter 8 will be concerned with the mechanisms by which the functional ends are achieved.

7.2 Competition and aggression

Competition is a term used by ecologists to describe the active demand by two or more individuals for a resource or requirement that is potentially limiting (Bakker, 1961; Wilson, 1975).

Nicholson (1954) distinguished between 'contest' competition and 'scramble' competition. Contest competition occurs when one individual obtains access to a supply of resources sufficient to maintain it, or enable it to reproduce, and it denies access to others. This form of competition includes not only aggressive behaviour but also some indirect methods such as reproductive competition (see below). Scramble competition involves each individual obtaining as much of the scarce resource as possible but without directly challenging the others. It occurs when the resource is widely distributed, and it differs from contest competition in that the less successful competitors are likely to obtain some of the resource.

Scramble competition occurs principally when fighting would be inappropriate as a means of securing a resource because it is widely distributed. In the case of food, the successful animal will be the one with the more efficient food finding and gathering mechanisms. In the case of male competition for receptive females, the successful animal will be the one which possesses skills to locate the potential mates (e.g. Wells, 1977, for anurans; Schwagmeyer & Woontner, 1986, for ground squirrels).

A second (and independent) distinction is between resource competition and sexual competition (Wilson, 1975). Resource competition occurs principally over food, but also over growing space (e.g. in sessile invertebrates), nesting materials and sites, and shade (Wilson, 1975). Sexual competition involves access to receptive mates: it includes both contest and scramble forms (see above). One form of sexual competition which is similar to scramble competition is unobtrusive mating, where a male sneaks up to one of a number of females which are being guarded by a male. This tactic is most successful when the dominant male is at the limit of its ability to guard the females, for example in elephant seals (Le Boeuf, 1974).

There are a number of indirect forms of sexual competition which fall into the category of 'contest' competition yet do not involve fighting. In males, competition may take the form of removing a rival's sperm prior to mating (in the damselfly: Waage, 1979), sperm competition (in insects, newts and chimpanzees: Parker, 1970; Verrall, 1984; Short, 1980), or olfactorily induced pregnancy-block (in mice: Bruce & Parrott, 1960). In females, it may take the form of suppression of the reproductive activity of other females.

We can conclude that whether a form of contest competition is to be regarded as aggression or not is principally a matter of *how* the competition is carried out, in particular whether it involves active interference, and also whether it occurs over short periods of time and is achieved by mechanisms specialised for harming or removing an opponent (see also Chapter 2).

7.3 Distribution of competitive aggression in the animal kingdom

Competitive aggression is widely distributed throughout the animal kingdom (Wilson, 1975; Huntingford & Turner, 1987). It has been relatively little studied other than in vertebrates and arthropods. Wherever aggression confers advantages over other forms of competition, and the animals possess effector organs and neurosensory mechanisms to exploit these advantages, we should expect it to have evolved. As we showed in Chapter 2, aggressive responses are found in sessile coelenterates such as corals (Lang, 1971, 1973) and sea anemones (Francis, 1973; Brace & Pavey, 1978). In these cases, competition is for growing space and for room on the substrate, respectively. Effector organs consist of a modified digestive response in the case of the corals and nematocyst-bearing structures (the acrorhagi: Fig. 2.1) used only for offence in sea anemones. Since the nervous system of the animals consists of a simple nerve net, mechanisms controlling the aggressive response will necessarily be simple.

Aggressive behaviour has been described in a few species from a wide variety of other invertebrate groups, such as nereid polychaetes (Evans, 1973), intertidal limpets and chitons (Stimson, 1970; Chelazzi *et al.*, 1983), nudibranch molluscs (Zack, 1975), cephalopods (Tinbergen, 1965), amphipod crustaceans (Connell, 1963), decapod crustaceans (Mitchell, 1976; Berrill & Arsenault, 1982; Dowds & Elwood, 1983) and arachnids (Riechert, 1978; Christenson & Goist, 1979; Austad, 1983). In most of these examples, the animals are competing for the occupation of space, whether on a rock or in a tube or on a web, and they all involve clearly recognisable aggressive interactions. For example, in the nereids, the proboscis is thrust forward, and the jaws may be used for biting (Evans, 1973); chitons dislodge the opponent by crawling over it and rocking backwards and forwards (Chelazzi *et al.*, 1983); and the limpet *Lottia gigantia* makes sudden thrusting movements with its shell to strike and push the intruder off the rock (Stimson, 1970).

Aggressive competition has been more widely studied in insects. Territorial behaviour and colony 'warfare' have been widely observed in the Hymenoptera (Wilson, 1975). Aggression between individuals within a colony can also occur, for example when a honeybee colony lacks both a queen and a brood capable of replacing her. In this case, the workers develop ovaries and show a dramatic increase in aggression at about the time of oviposition (Sakagami, 1954).

Competitive aggression is much better known and has been more extensively studied in vertebrates, particularly in teleosts, birds and mammals, on which most field and laboratory studies have been carried out (Wilson, 1975, Chapter 12). In these animals, there are many examples of weapons, specialised effector organs for fighting (e.g. Geist, 1966) and the neural mechanisms controlling fighting also show specialisation (Moyer, 1968).

7.4 A functional analysis of interspecific differences in aggressive competition

The evolutionary rules underlying interspecific variations in competitive aggression for food resources have been well covered in a number of previous discussions, for example those of Brown (1964), Wilson (1975), Clutton-Brock & Harvey (1976) and Geist (1978). When food is abundant, aggression will be unnecessary since the same benefits can be obtained without it. When food is scarce, it will often be advantageous for the animal to use its energy in foraging for food (i.e. scramble competition) rather than in aggressive competition. This will apply particularly when food is widely dispersed or difficult to find. In general, therefore, we might expect aggression to occur under conditions of intermediate food availability.

The wild boar and the reed buik provide examples of animals which show low rates and severity of aggression as a consequence of inhabiting productive environments (Geist, 1978). The warthog and the muntjac both live in dry, low-productive, habitats, and show low levels of aggressiveness. Other studies have shown decreased aggression within a single species as a result of the habitat becoming unproductive, and food becoming widely distributed and scarce (in rabbits: Mykytowycz, 1961; in chacma baboons: Hall, 1963; in rhesus monkeys: Southwick, 1969, and Loy, 1970).

If it is advantageous for an animal to compete aggressively for food, it will be energetically more efficient for it to fight a relatively few times for an area which contains food than to compete for each item of food (Wilson, 1975; see also Chapter 2). Given such a broad generalisation, we should expect all animals to seek to defend feeding territories either in groups, in pairs, or individually, unless it is not possible to defend their food supply in this way. This leads us to the question of the circumstances in which resources are defensible.

The answer has been sought both in general surveys of the relationship between social structure and ecology, associated with the socioecological approach initiated by Crook (Chapter 1), and also in the more formal 'optimality' models of territory economics (see also Chapter 1).

Geist (1974a) surveyed the relationship between social organisation and ecology in ungulates and concluded that territoriality had evolved in habitats where their plant food was plentiful and was renewed continuously. If food was either unpredictable and dispersed, or localised and highly abundant, territoriality was absent, in the first case because the food could not be defended and in the second because it was not in short supply. In a study of the social structure of antelopes, Jarman (1974) identified five basic types of organisation and concluded that individual territories were found in species which browsed on high-quality food even though it was relatively scattered. Animals which fed off poorer-quality, scattered food, on the other hand, moved around in large herds following the growth cycles of the pastures, as in the case of the wildebeest.

J. L. Brown (1964) originated the economic approach to territorial behaviour when he put forward the idea of 'economic defendability' (Chapter 1). Territorial aggression was viewed in terms of resource competition, but whether a resource was defended in a particular case would depend on the costs of defence in relation to the benefits obtained. Benefits include food availability, acquisition of mates, a safe place to rear young and, in some cases, predator avoidance (Brown, 1964). Costs include time and energy spent in defence and in foraging, risk of injury and predation (Davies & Houston, 1984; Huntingford, 1984a). Brown predicted that territoriality would have evolved only when the benefits exceeded the cost. However, since such diverse factors are involved in both cases, it is difficult to establish a common currency to test this prediction with a wide range of resources. In practice, empirical investigations have been largely restricted to feeding territories where benefit can be quantified in terms of energy intake (Davies, 1978a; Huntingford, 1984a). The costs which have been measured include territorial defence (Gill & Wolf, 1975; Riechert, 1979; Frost & Frost, 1980; Puckett & Dill, 1985), predation pressure (Horn, 1968), travelling cost of food foraging (Horn, 1968; Puckett & Dill, 1985; see also Andersson, 1978), and reduced energy intake resulting from territorial interference (Jaegar, Nishikawa & Barnard, 1983).

Nevertheless, there is still a problem of what value is being maximised: is it, for example, rate of food intake, or some long-term goal such as survival of the nestlings in the next breeding season (Davies & Houston, 1984)? The only practical solution to this problem is to test models incorporating different optimisation criteria against measures taken in the field to see which one provides the best fit.

One of the earliest empirical studies was that of Horn (1968) on Brewer's blackbird. He derived a qualitative model, from Brown's principle of economic defendability, which predicted that territorial defence would occur only when the food resource was stable (i.e. it was renewed at a constant rate) and evenly distributed throughout the habitat. Only then would the travelling costs be sufficiently low to make territorial defence advantageous. If food were unpredictable and concentrated in patches, the costs of food foraging within a territory would be too high, and it would be more efficient to forage from a central colony. Incidentally, it is interesting to note the parallels between these predictions and the results obtained from socioecological surveys (e.g. Geist, 1974a; Jarman, 1974). Horn calculated the foraging pattern for Brewer's blackbird, a colonial nesting bird, and found that it did, indeed, conform to the pattern he predicted for colonial nesting. However, no comparable calculation was made for a territorial species, and Davies (1978a) pointed out that Horn's model takes account of only one cost, that of food foraging, and is therefore of limited applicability.

So far, we have considered interspecific variations in territorial behaviour. A more quantitative model based on the cost–benefit approach was used by Carpenter & Macmillan (1976) to predict when territorial behaviour would occur within a single species, the Hawaiian honeycreeper, a nectar-feeding bird. The model predicted that the birds would defend territories at intermediate flower abundances, but not when these were low or high. Empirical findings did, in general, support the model's predictions of when territorial behaviour would occur, and how intense it would be, although the minimum number of flowers necessary for establishing and maintaining territories was higher than predicted. This finding raises again the issue of the time period over which an adjustment in accordance with cost–benefit requirements is being made. As Carpenter & Macmillan pointed out, the adaptive 'goal' of territoriality may be to provide a guaranteed food supply over a length of time, rather than to respond to changes over shorter periods of time.

In another study of a nectar-feeder, the golden-winged sunbird, Gill & Wolf (1975) did find that fluctuations in territorial behaviour appeared to correspond to short-term changes in nectar levels. A later analysis of their data by Pyke (1979) assessed it in terms of four alternative theoretical models each containing a different fitness criterion or 'currency'. A model whose criterion was minimisation of daily energy costs, subject to the constraint that sufficient energy must be obtained to stay alive, fitted the data well, whereas a model with maximisation of energy intake as the criterion did not.

These studies raise two general issues involved in the construction of optimality models of territorial defence. One concerns the criterion choice, and the other (which is related to it) concerns the time period

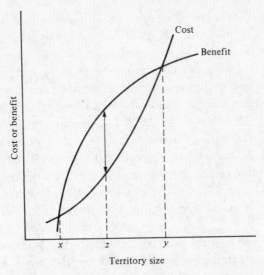

Fig. 7.1. Hypothetical relationship between costs and benefits and territory size: x is the threshold at which territory defence becomes economic; y is the threshold beyond which it becomes uneconomic; z is the size where maximum benefit is obtained. (Based on Davies & Houston (1984), Fig. 6.1*a*.)

over which this criterion is maximised. Both are discussed by Davies & Houston (1984). With regard to the time-scale of adaptive changes, they contrasted three bird species for which data are now available. Nectar–feeders, such as the sunbirds referred to above, defend territories for short periods and abandon them once they have become unprofitable. Pied wagtails, on the other hand, defend territories throughout the winter, even on days when alternatives would be more profitable (see below). Even more stable in their behaviour are ural owls, which spend their whole lives on the same territory, and will defend them even when prey are so scarce that there is no chance of breeding.

These three examples show that the occurrence of territorial defence is sensitive to different maximisation criteria in different cases. They also suggest that territorial aggression can be controlled by mechanisms with different degrees of flexibility. We return to this point in the next chapter when we consider the mechanisms underlying competitive aggression.

The principle of economic defendability has also been applied to territory size (Davies & Houston, 1984). The basic assumption is that both costs and benefits will increase with size, but the curve for benefits

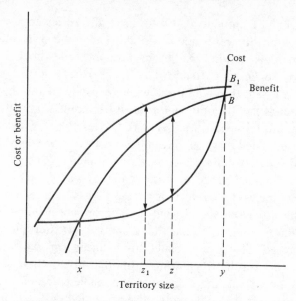

Fig. 7.2. Hypothetical relationship between costs and benefits, and territory size, showing that a higher benefit (B_1 instead of B) leads to a new value of z (z_1 instead of z) where maximum benefit is obtained. (Based on Davies & Houston (1984), Fig. 6.1.*b*.)

will be asymptotic (since resources can become superabundant), whereas that for costs will be exponential (since defence costs may rise steeply with additional intruders). In Fig. 7.1, beyond point y, territorial defence will be uneconomic, whereas at point z it will lead to maximal benefits.

Fig. 7.2 shows the same graph with an additional curve for benefits (B_1) that represents a higher value (e.g. when more food is available): in this case, the value of z will be moved to z_1, where territory size is smaller. Davies & Houston (1984) have reviewed two detailed studies of bird species, which show that territory size is adjusted to changes in resources as predicted by this graph. Rufous hummingbirds respond to removal of flowers from their territory by expanding the territory size to encompass a constant number of flowers (Hixon *et al.*, 1983). They also adjust their territory size to take account of the different nectar content of different flower species. Several studies of sanderlings, a shorebird, have shown that their territories are, as predicted, smaller when resources are more abundant. The mechanism underlying this adjustment appears to involve a response to intruder density rather than an adjustment to the amount of food in the territory.

Davies (1978*a*) and Huntingford (1984*a*) have also reviewed a wider variety of studies on different animals to show that the principle of economic defendability can, in broad terms, account for territory size.

In contrast to the findings described so far, territory size in pied wagtails is held constant throughout the winter, despite the prediction on economic grounds that the optimal territory size should vary with the rate of food renewal (Houston, McCleery & Davies, 1985). Davies & Houston (1984) suggested that continual adjustments may be costly, and that territorial defence is more a long-term investment in this species: when food was very scarce on the territories and the owners had to leave to feed, they nevertheless kept returning to re-establish their territories. This brings us back to the issue of what the appropriate maximisation currency is: in the wagtail, it is clear that a short-term criterion is not involved, but instead behaviour is related to a longer-term goal such as maximising survival over the whole winter. In other cases, where short-term criteria are important, there are several possible currencies, such as time minimisation (in the golden-winged sunbird: Pyke, 1969), or the rate of weight gain or energy maximisation (in rufous hummingbirds: Carpenter, Paton & Hixon, 1983; Hixon *et al.*, 1983). Schoener (1983) provides a more detailed discussion of these possible currencies.

So far, we have considered forms of competitive aggression based on the defence of a spatial area where food is located. There are other forms which competitive aggression for food can take, principally among members of a group of animals which occupy the same area. In such groups, food (or another resource) may be irregularly distributed and animals may compete aggressively for the resource. This type of aggression is complicated by two features. First, there are circumstances where prior 'ownership' of the resource will confer an advantage, i.e. simply being in possession will decide access to the resource. The reasoning behind this is based on a game theory analysis which will be described in Chapter 9. The second feature, dominance, has been more widely recognised in the study of animal societies and in comparative psychology (Chapter 1).

Priority of access to resources within a group of animals is generally established as a result of the outcome of previous aggressive encounters, which includes the initial encounter and subsequent fights or threats.

There is no problem in accounting for the behaviour of the dominant animal in functional terms. To gain access to scarce resources without having to fight on each occasion is clearly advantageous in that time, and energetic and injury risk costs are minimised (Wilson, 1975). However, we must also consider the subordinates. In many species, these can be recognised by characteristics such as their behaviour and posture, odour or even visual characteristics (e.g. Rowell, 1974; Wilson, 1975; Rohwer & Rohwer, 1978). Size and age are frequent correlates of dominance in the animal kingdom (see Wilson, 1975, table 13–2), so that the first possible adaptive feature of being subordinate is to avoid potentially damaging encounters at an age when there is little chance of winning them. If such encounters do not concern resources necessary for immediate survival, it may be advantageous to avoid contesting them. As Wilson (1975) pointed out, there is a high turnover in dominance status for animals such as the black grouse, red deer and many primates, and this turnover is age-related (see also Clutton-Brock *et al.*, 1979, 1982, for the red deer; Reiter *et al.*, 1981, for the northern elephant seal; Wilson & Francklin, 1985, for the Chilean guanaco; and, Alados, 1986, for the Spanish ibex).

In the case of sexual competition (see section 7.5), another possible adaptive feature of subordinate behaviour is evident in those species where alternative mating strategies are found. Here, subordinates may occupy a peripheral position and gain access to a usually guarded mate when the dominant animal is distracted. Nevertheless, even in the absence of age-related dominance or alternative tactics, it may still pay a subordinate to use suboptimal resources in safety rather than risk a damaging fight with a dominant animal which it has little chance of winning.

7.5 Sex differences

In many accounts of animal aggression, the male is seen as the more aggressive sex. This is usually attributed to intermale competition for access to females, one of the two components of Darwin's (1871) theory of sexual selection, the other being female choice.

In this section, we first consider explanations derived from Darwin's sexual selection theory, and then question whether the common view of the more aggressive male is an accurate one. This leads us to consider additional explanations to that of male competition.

One of the most influential contemporary functional theories of sex differences is that of Trivers (1972), which we encountered briefly in Chapter 5 in relation to parental aggression. Essentially, Trivers sought to explain why sexual selection usually takes the form of male competition and female choice. He argued that the female generally invests more than the male in her offspring, because the greater initial investment in the gametes by the female makes it more advantageous for the male to leave and to seek further females, rather than caring for the young. The consequence of this imbalance would be for the female to become a limiting resource and for males to compete for access to them. Trivers argued that this would be the basic pattern wherever there was anisogamy, unless the male invests more in the offspring through some other means,* as occurs in polyandrous wading birds where male parental investment is higher than that of females (Jenni, 1974; Harding, 1983), or where both parents are needed to help feed and defend the offspring, as occurs widely in monogamous birds.

Whenever there is a great imbalance in parental investment, either by the male making little or no contribution to parental care, or by the reverse case in polyandrous species, the reproductive success of the individuals of the limited sex will potentially be very high. However, it will be subject to restrictions imposed by access to the limiting sex (such as female choice), and by intrasexual competition.

Trivers' theory explains sex differences in features associated with aggressive competition, such as size, fighting ability and weapons. It also accounts for the association between pronounced intermale competition and little or no male parental investment, as occurs in species such as the elephant seal, where males are much larger than females, competition is pronounced and a successful male can hold a one-male social group (cf. Gowaty, 1980) containing about 40 females (Le Boeuf, 1974). It also explains why females from polyandrous species show typically 'masculine' features, such as bright plumage, larger size and aggressiveness; and why there is little intermale competition, together with sexual dimorphism in size, aggressiveness and weapons, accompanying biparental care. But the model leaves the following features unexplained:

1 the origin of parental investment patterns;
2 the form of male competition, whether it occurs through fighting or an alternative means;
3 the possible contribution of features unrelated to sexual selection, such as between-sex competition, and female competition.

Trivers' model is concerned only with the *consequences* of parental investment patterns. How these arise (other than through the initial

---

*Although there was a flaw in Trivers' original logic of linking past investment to future prospects (the 'concorde fallacy': Dawkins & Carlisle, 1976; see also Chapter 5), the practical limitations imposed by this are few, since past investment is usually a good indication of the future investment needed to produce offspring to the stage they were at when they were lost.

imbalance in gametic investment) is not considered. To do so requires consideration of prior adaptations and the constraints these impose on the subsequent course of evolution (Gould & Lewontin, 1979). For example, the combination of territoriality with external fertilisation in teleosts can be viewed as a preadaptation for male parental care (Chapter 5: see Williams, 1975). In mammals, the combination of internal embryonic development, placental and lactational nutrition is a constraint which predisposes the majority of species to maternal or biparental care.

Ecological conditions, such as how food or other resources are distributed, and the temporal availability of receptive mates, will also affect the pattern of sexual dimorphism (Emlen & Oring, 1977). Where resources are clumped together, a few animals may be able to monopolise them and acquire several mates. Where resources are widely distributed, a monogamous mating system is more likely (see Fricke, 1980; Davies & Lundberg, 1984; Davies, 1985; for studies showing intraspecific variations in the mating system as a consequence of different resource distributions).

If female receptivity is synchronous, the equivalent of a temporal 'clumping' of receptive mates will occur, and the cost of defending the limiting sex will be lower, enabling polygyny based on defence of a one-male social group (Gowaty, 1980), or on male dominance, to occur (Emlen & Oring, 1977).

Thus, spatial distribution of resources and temporal availability of mates will affect the cost of defending either the resources or the mates. The benefits depend on the requirements of parental care. Where both parents are required to ensure the young's survival, males will not benefit from polygyny. Where only one parent is required, polygyny will be more beneficial to the male (Emlen & Oring, 1977).

Considering the form male competition may take, there are a number of alternatives which do not involve aggression (section 7.2), such as sperm competition and sexual interference. Male aggression itself takes many forms, including direct interruption of copulating pairs (e.g. Sivinski, 1978; Berrill & Arsenault, 1984), competing for territories or leks to which the female is attracted, and guarding the female immediately before or after copulation (in insects, ungulates and primates: Parker, 1974a; in crustacea: Berrill & Arsenault, 1982; and possibly in rats: Flannelly *et al.*, 1982).

Male competition is further complicated by the existence of alternative tactics within species (Austad & Howard, 1984). In many species, including crickets, hanging flies, treefrogs, iguanas and elephant seals, males commonly use more than one method of gaining access to females (Huntingford, 1984*a*). Perhaps the best-known example occurs in the ruff (*Philomachus pugnax*), where there are independent and satellite males which differ in plumage, the former being dark-coloured and the latter paler and white (Van Rhijn, 1974). Independent ruffs show more fighting, and they can be subdivided into two types, residents, which defend territories, and marginals, which do not. Satellites do not defend territories but have access to them.

Van Rhijn suggested that the resident and the satellites derive a mutual benefit from one another's behaviour: the resident male cannot usually attract enough females for copulation by itself, but the presence of satellites increases the attractiveness of the territorial area; the satellites derive most benefit from being able to copulate early in the season when the residents tend to attack one another.

In salmon such as *Oncorhynchus* spp., some males ('jacks') mature at a much younger age than others and attain only a fraction of the body size of later maturers ('hooknose'). These two forms gain access to females, respectively, by sneaking and fighting. Gross (1985) calculated that the lifetime fitness of the two tactics would be roughly equal and that intermediate forms would be at a disadvantage. He therefore argued that they represent a population at a 'mixed evolutionarily stable strategy' (Maynard Smith, 1982): any increase in the proportion of jacks would force them to fight other males, consequently reducing their access to females, and their subsequent numbers; any decrease in jack frequency would increase the number of sneak matings, and hence increase their subsequent numbers.

Alternative tactics take advantage of one of the costs of fighting over a resource, namely that it will be unguarded while the fight is progressing and the combatants will need to attend to each other if they are to fight efficiently. We should therefore expect these tactics to occur where fights are more frequent or prolonged, and where the 'resource' (the female) is not readily defensible.

So far, our discussion has been based on sexual selection or extensions of it. Hrdy (1981) argued that Darwin's analysis, by its concentration on male competition and female choice, had neglected the evolution of female traits. To redress this balance, we have to consider selective forces which may affect the two sexes in different ways (Hrdy, 1981; Floody, 1983). One of these is the extent to which the two sexes actively compete with, or avoid competing with, individuals of the opposite sex, for resources such as food. Another is aggressive competition between females.

Overt aggression between the two sexes certainly occurs in some species. The male may come to dominate the female (e.g. in the nectar-feeding bellbird: Craig & Douglas, 1984; in Mexican jays: Barkan *et al.*, 1986), or the female may dominate the male (e.g. in the spotted hyaena: Hamilton, Tilson & Frank, 1986; in non-breeding reindeer: Bouissou, 1983; in the golden hamster: Payne & Swanson, 1970; in spiny mice: Porter, 1976; and in other mammals: Hrdy, 1981; Floody, 1983). Females may also drive males away from an area they occupy (e.g. in the mountain goat: Geist, 1978; the European vole: Frank, 1957; the golden hamster: Lisk *et al.*, 1983).

Against these cases, we have to balance widespread examples where the two sexes avoid aggressive interactions by living in different social worlds (Hrdy, 1981, for primates, following Wrangham, 1980; Clutton-Brock *et al.*, 1982, for cervids). Floody (1983) also concluded that competitive aggression between adult males and females may be infrequent in mammals. Experimental studies from a wider range of animal groups provide examples of males and females forming separate dominance hierarchies (e.g. green swordtail fish: Beaugrand, Caron & Comeau, 1984), or showing individual or group territorial aggression mainly to their own sex (e.g. the stone crab: Sinclair, 1977; the Norway rat: D. C. Blanchard *et al.*, 1984; De Bold & Miczek, 1984; Hood, 1984; the housemouse: Butler, 1980; Ebert, 1983; the California ground squirrel: Owings, Borchert & Virginia, 1977; the rhesus monkey: Southwick, 1969; the common marmoset: Evans, 1983).

Absence of competition between the sexes may extend even further where they show specialisations for different ways of life. Selander (1972), following Darwin (1871), suggested that some sexually dimorphic features in birds are the result of adaptations to different feeding habits. He argued that ecological specialisation could lead to dimorphism in body size, and suggested this as an explanation for the 'reversed' sexual dimorphism which has evolved independently in several groups of predatory birds, where males and females take different sizes and types of prey. A similar argument was advanced by Brown & Lasiewski (1972) to explain the sexual dimorphism of elongated carnivores such as the weasel (cf. niche specialisation in some species of dioecious plants: Freeman, Klikoff & Harper, 1976).

Competitive aggression between females is also neglected by the sexual selection view. Its function may be to provide access to the opposite sex, for example by preventing another female from coming near to a territorial male (in the tree lizard, *Liolaemus tenius*: Manzur & Fuentes, 1979) or to a male which will then provision young (e.g.

Yasukawa & Searcy, 1982, for the red-winged blackbird). It is more common, however, to find that female aggression results in the occupation of feeding territories (e.g. in cervids: Clutton-Brock *et al.*, 1982; in anurans: Wells, 1977; in primates: Wrangham, 1980). In a wide range of animals, from the social Hymenoptera (Breed & Bell, 1983) to primates (Hrdy, 1981), reproductive suppression of subordinates is another consequence of female aggression.

7.6 Life cycle changes

In this section, we consider changes throughout the life cycle. We examine competitive aggression in young animals, first in relation to a classification of life histories, and secondly in terms of its developmental consequences. We then trace the development of competitive aggression across the life span and briefly discuss changes with age from a functional perspective.

Competitive aggression in young animals can be related to a more general issue in population ecology, interspecific differences in adaptive patterns of life history, notably the distinction between *r*-selected and *K*-selected species (Horn & Rubenstein, 1984). *r*-selected species show a high reproductive rate, a short generation time, small size and fluctuating birth and death rates. *K*-selected species, on the other hand, show a low reproductive rate, a long life span, large size and steady population numbers. In general, a typical *r*-selected species would be a small invertebrate, and a typical *K*-selected species a large bird or mammal.

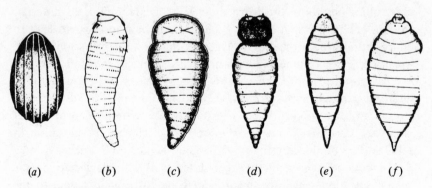

(*a*)   (*b*)   (*c*)   (*d*)   (*e*)   (*f*)

Fig. 7.3. Immature stages of the parasitic wasp *Poecilogonalos thwaitesii* showing transformation from the egg (*a*) to successive larval forms including the fighting form (*d*) which possesses a sclerotised head and large mandibles. (From Clausen (1941), p. 60.)

Although this distinction provides a highly simplified picture (Horn & Rubenstein, 1984), it can provide a useful framework for considering when aggressive behaviour first appears in development. *r*-selected species generally show rapid development, but there may be intense competition early in life since mortality is very high. The young of *K*-selected species show slower development and are more sheltered by their parents. Furthermore, most large animals delay reproducing until they grow to a certain size, and since much competitive fighting is associated with the adult reproductive state, we should expect delayed development of the adult pattern of aggression in these species.

Most examples of the early appearance of adult-like patterns of aggressive behaviour are indeed found in *r*-selected species. In some parasitic Hymenoptera, the larvae undergo a transformation into a fighting form that kills and eats conspecific larvae occupying the same host (Figs 7.3 and 7.4). The larvae of the caddis-fly are predatory net-spinners, and they engage in agonistic encounters over the possession of nets (Hildrew & Townsend, 1980). Aquatic larvae, such as those of the dragonfly may also be territorial (Harvey & Corbet, 1985, 1986) and success as a larva enhances adult mating success. Olomon, Breed & Bell

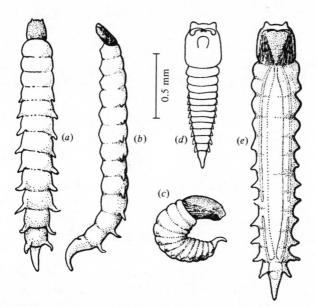

Fig. 7.4. Immature stages of the parasitic wasp *Collyria calcitrator* showing progression from the first instar (*a–b*) to the second instar (*c–e*) including the fighting form (*b*) which possesses a sclerotised head and large mandibles. (From Clausen (1941), p. 87.)

(1976) found that the entire agonistic repertoire of cockroaches was shown by the nymphs.

Young teleosts have been observed to show aggressive behaviour. In juvenile salmon, for example, territorial behaviour occurs in the freshwater (parr) stage (Villars, 1983). Brown & Colgan (1985) showed that aggressive behaviour matured at different ages in the fry of four species of fish, and the timing corresponded closely to ecological changes such as the departure of the guarding parent. In reptiles, competitive aggression also occurs early in life. The iguanid lizard, *Sceloporus jarrovi*, establishes a territory within weeks of hatching, and it has even been claimed that *Anolis aeneus*, another species of lizard, forms territories almost at birth (Burghardt, 1978).

Among *K*-selected species, the majority of birds and mammals show a period of dependency on their parents which generally precludes the necessity for early competitive aggression on the scale found in *r*-selected species. However, there are some exceptions, for example in certain raptors where aggression among chicks is carried to the extreme of the elder of two siblings killing the younger. This is probably a method of brood reduction permitting the parents to rear the maximum numbers of young that can be fed (Stinson, 1979). More generally, precocial birds such as domestic fowl chicks will show aggressive pecks as early as 1 day after hatching (Rajecki, Ivins & Rein, 1978), and fighting for food has been observed in young birds (Bekoff, 1981).

In mammals, the young of species such as the hyrax and the pig have been observed to fight for access to milk. Geist (1978) has described the short, sharp, but temporary canine teeth that are used by suckling pigs in encounters which determine the position the animal will occupy on the maternal teats. The preferred position is further up the row, and a pig occupying such a position will be larger, and show advantages as an adult.

Bekoff (1981) has pointed out that success in aggressive encounters early in life may have cumulative effects only apparent later, and he explained the high level of aggression found among juvenile lizards (see above) in terms of its influence on the acquisition of future rather than present resources.

Geist (1978) has suggested that in social mammals, such as canids and some ungulates, rigid dominance relations are established early in life and are stable throughout adulthood (there is similar evidence for developmental continuity in humans: Olweus, 1979, 1984). Geist also suggested that play fighting is a form of 'gaining dominance', and hence would provide a long-term fitness benefit. However, Einon (1983) reviewed animal and human studies to show that rough-and-tumble play is related neither to dominance relations nor to individual differences in aggression. Einon (1980) had argued that the distribution of rough-and tumble play in mammals (in primates, carnivores and social rodents) was inconsistent with it being closely related functionally to aggression, but supported the view that its main beneficial consequence is increased flexibility and social responding. (But see Martin & Caro, 1985, for a review which questions the evidence for play conferring important long-term benefits.) In contrast to 'real' fighting, rough-and-tumble play is characterised by rapid reversal of roles, is frequently preceded by an invitation signal, and damaging acts such as biting are absent (Einon, 1983).

Despite the causal and functional separation of rough-and-tumble play and adult fighting, they merge into one another developmentally in several species including the polecat, the Norway rat, the Californian ground squirrel and several canids (Einon, 1983). Einon has suggested that adolescent fights are distinct from both rough-and tumble play and from adult fights. In those rodents which do not show rough-and-tumble play, such as the housemouse, adolescents show low-intensity fights, which can be distinguished from rough-and-tumble play because they tend to be connected with disputes over a resource and the chases are not reciprocated. Hole & Einon (1984) have classified these as a second form of play fighting, designated by the clumsy acronym NVSF ('not very serious fighting').

Meaney & Stewart (1981), investigating the social development of rats, found that rough-and-tumble play became increasingly similar to adult fighting around the time of sexual maturity. In Hole & Einon's terminology, rough-and-tumble play is becoming increasingly replaced by NVSF, which is more clearly related to adult fighting. Viewed in this light, Geist's (1978) suggestion that play fighting is important in some species for achieving long-term benefits associated with dominance may after all be correct for NVSF (but not for rough-and-tumble play), although we should note the general doubts about the evidence that play confers long-term benefits (Martin & Caro, 1985).

Taylor (1980) found a high degree of developmental continuity between individuals in what he called their 'juvenile' fighting, and he suggested that play fighting might be a misleading term for this. By distinguishing such juvenile fighting or NVSF from rough-and-tumble play, we can reconcile Taylor's results with those from studies showing that rough-and-tumble play is causally and functionally different from adult fighting (Einon, 1980, 1983; Einon & Hole, 1984).

The emergence of the adult pattern of aggression at the time of sexual maturation in K-selected species such as birds and mammals (Brain, 1977, 1979) makes sense in terms of restricting potentially damaging and costly encounters to the time when the resources obtained can be used for reproduction. In species where the female is characteristically more aggressive than the male, the adult pattern also develops at the time of sexual maturation (e.g. in the golden hamster: Goldman & Swanson, 1975). Comparable changes in aggressive behaviour, coinciding with sexual maturation, have been found in some teleosts (e.g. in the paradise fish: Davis & Kassel, 1975).

In many non-tropical animals, breeding occurs seasonally, and there is likewise a restriction of potentially damaging and costly encounters to the reproductive season. In many teleosts, birds and mammals, the male holds a territory or shows intrasexual aggression only within the breeding season, coinciding with the rise in testicular activity (e.g. Bouissou, 1983; Harding, 1983; Villars, 1983). In species where females show territorial defence, this may also coincide with the breeding season (e.g. in the bank vole: Bujalska, 1973), although little attention has been directed towards seasonal changes in female mammals (Floody, 1983).

The females of many species show changes in aggression across their reproductive cycle. Some of these were discussed in the chapter on parental aggression, but others occur prior to fertilisation. Most research on this topic has been carried out in relation to the oestrous cycle of rodents. The evidence suggests that changes in competitive aggression occur across the oestrous cycle in some species but not in others (Floody, 1983).

The female hamster shows heightened aggressiveness towards males prior to sexual receptivity, and this can be viewed functionally in terms of securing a burrow or territory (Lisk *et al.*, 1983). Once the female has established a burrow, the main competition will come from other females, and this is reflected in a generally high level of aggression towards female intruders (Takahashi & Lisk, 1983, 1984).

Floody (1983) speculated that in species which are induced ovulators, and therefore can be in oestrus for some time, it would be disadvantageous to run the risk of prolonged lowering of aggression at oestrus. The collared lemming, which shows little change over the cycle, fits this prediction. Among induced ovulators, where the female is normally highly aggressive when not in oestrus (such as the hamster), Floody suggested that a significant decrease in aggression during the (short) oestrous period would be advantageous. However, it is apparent from the studies described above that any lowering of aggression at oestrus is sex-specific in terms of its target.

Floody also suggested that in species where the female is normally less aggressive than the male, such as the housemouse and the rat, changes similar to those found in the hamster (but less pronounced) would occur. There is some evidence that this is so. The housemouse shows lower aggression towards female intruders at oestrus and dioestrus than at proestrus and metoestrus (Hyde & Sawyer, 1977). In the laboratory rat, aggression towards male and female colony intruders was found to be lowest at oestrus but greatest a day later, at dioestrus (Barfield, 1984; Hood, 1984).

Bouissou (1983) reviewed evidence for oestrous cycle changes in ungulates, and concluded that *increased* aggressiveness occurs in some species at oestrus, for example in ewes, horses, cattle and one-humped camels. Of course, this aggression is directed towards other females and it can be viewed as broadly similar to the interfemale aggression of hamsters, as a form of intrasexual competition.

It is clear from these studies that it is crucially important to consider the sex of the opponent in seeking to explain the pattern of change over the oestrous cycle.

So far, we have considered aggression in young age-groups, its long-term developmental consequences, and how competitive aggression is affected by reproductive condition in vertebrates. In some invertebrates, other life-span developmental factors, for example the moult cycle, may be important. Crustacea such as the American lobster show increased aggression as they approach the moult (Breed & Bell, 1983). In functional terms, this serves to space them out at a time when they become vulnerable.

Considering changes with age, two contradictory predictions can be derived from two different functional models. Trivers' (1972) parental investment model predicts that as an animal becomes older, it should be more prepared to take risks in aggressive encounters since its potential to reproduce (reproductive value: Fisher, 1930) will be lower (cf. Andersson *et al.*, 1980, for parental aggression). There is, however, very little evidence on this point. Clutton-Brock & Harvey (1976) supported the prediction by citing a study of chimpanzees which found that older individuals may be more prepared to contest access to meat supplies than younger ones. In a study of the Spanish ibex, Alados (1986) reported that older males expended more energy during the rutting season than younger ones.

On the other hand, Parker (1974*b*) predicted that aggressive encounters should decline as an animal becomes older because it will perceive that its chances of winning are reduced, and it will be less likely to attack. While Trivers considered only increased willingness to take risks in relation to expected lifetime reproductive success, Parker considered only decreased willingness to take risks in relation to the probability of injury. A model which seeks to balance both aspects is necessary in future.

One study of laboratory rats (R. J. Blanchard *et al.*, 1984*a*) found that the level of male aggression in mixed-sex colonies formed at different ages showed no decline with age (up to approximately 2 years of age); and Fairbanks & McGuire (1986) found no evidence for reduced involvement in aggressive encounters with age in groups of female vervet monkeys.

7.7 Conclusions and suggestions for future research

I began this chapter by placing competitive aggression into the wider context of competing for potentially limiting resources. Aggressive competition involves active interference, it occurs over short time periods and is often achieved by specialised mechanisms. The discussion of when food resources would be subject to aggressive competition concentrated on two issues: first, when animals would defend feeding territories rather than individual food items, and secondly, dominance relations. Research on territory economics, based on Brown's principle of economic defendability, has involved studying variations within a single species, such as nectar-feeding birds. Changes from territorial to non-territorial behaviour, and territory size, have both been shown to vary with short-term fluctuations in food levels. However, in birds such as pied wagtails, territory size was maintained at a constant level, despite short-term fluctuations, suggesting a longer-term fitness criterion is involved. Considering dominance, the benefit for a dominant animal would be that resources need not be contested anew each time, and for the subordinates that potentially damaging encounters were avoided.

Sex differences were considered both in relation to sexual selection theory, which emphasises male competition, and also in relation to other issues ignored by this approach, such as intersexual competition and female competition.

Concerning life cycle changes, it was suggested that an adult-like form of competitive aggression appears early in life in *r*-selected species;

whereas in *K*-selected species there is a delay in its appearance. For several species, there was evidence that aggression occurring early in life conferred benefits in adulthood. In *K*-selected species, the onset of pronounced competitive aggression usually coincides with sexual maturation, or is restricted to the breeding season.

Further research is required on all the main issues covered in this chapter. The economic rules underlying feeding territories have been studied in only a small number of species, and have been largely restricted to feeding territories. Sexual competition has been studied mainly within the framework of sexual selection theory. There is also relatively little systematic evidence on the timing of the appearance of aggression in development. Some issues, such as the developmental continuity of aggression, and its relation to rough-and-tumble play in mammals, have been more extensively investigated, although further clarification is necessary.

Rather than list specific suggestions for research, it is perhaps more useful to make some general remarks about the functional models covered in this chapter. Economic models are based on the likely costs and benefits of behaviour. In order to make them empirically testable, they need to be concerned with as few variables as possible. Thus, the territory models, which have generated the most useful empirical work, have concerned the costs of defence and the benefits of energy intake. Sexual selection models again involve a relatively few variables (such as parental investment), which, although less precise than those used in territory models, can still be operationalised.

This approach parallels the traditional reductionist method which has been used successfully in other fields of biology. It entails investigating important features one by one, and of choosing simple examples where additional variables and interactions with other systems are absent (Waddington, 1977). Such additional variable are often referred to as 'constraints'. They include environmental constraints, which result from trade-offs with features not considered in the specific model, and internal constraints, concerned with the way the mechanisms underlying behaviour operate. These are determined by past evolutionary history, and also by what Gould & Lewontin (1979) referred to as architectural constraints, consequences of the overall design of the animal.

In relation to a model of territorial behaviour in nectar-feeding birds, Pyke (1979) calculated that minimisation of daily energy costs accounted for changes in territorial defence, but this was subject to the 'constraint' that sufficient energy must be obtained to stay alive. The issue of what determined this value was beyond the scope of the specific model, and could have been related to either of the two types of constraint described above. In a study of two related species of colobus monkeys, living in a similar general environment, Clutton-Brock (1974) found that the diet of the two species was very different, and this led to differences in their microenvironment and in their social structure. Here, diet may have been determined by constraints on the digestive system, presumably derived from past evolutionary history or architectural constraints.

The criticism that evolutionary biologists ignore constraints and concentrate on immediate adaptations to local conditions was the essence of the attack on the functional viewpoint made by Gould & Lewontin (1979). They argued that constraints restrict the possible paths and modes of change so strongly that these constraints themselves are the most interesting aspects in determining the path of evolutionary change.

These critics are generally regarded as being unnecessarily destructive and as underestimating the undoubted achievements of the functional approach. Nevertheless, they did indicate a weakness of the economic approach, which its more cautious adherents recognised all along, and which is now becoming more fully addressed in research.

Eventually, a satisfactory functional approach will have to encompass not only constraints imposed by the environment and by the animal's mechanisms but also a consideration of how the costs and benefits of different forms of behaviour trade off with one another. In terms of research strategies, this represents a switch from the reductionist approach of considering simple examples to a systems approach which incorporates the interaction of a large number of variables.

Such an approach was outlined by Vehrencamp & Bradbury (1984) in relation to mating systems. They advocated adopting a broad interactive view. Nine different components of fitness which affect an animal's overall mating system were listed. These included the numbers of females a male encounters and courts within a season, the survival probability of the offspring and adult male survival rate. For each of the nine fitness components, there were various behavioural options which would enhance that component, and the overall mating system was seen as an integrated subset from all the possible options. Anatomical and physiological constraints, together with the adaptive requirements of the present environment, will set the limits to the options available in any one case. Within these limits, the behavioural options that combine together to produce the overall mating systems can, according to Vehrencamp & Bradbury, be best studied by an extension of the game theory approach. This would, however, need to take more account of issues such as constraints and alternative adaptative pathways than do many existing game theory models.

Further reading

Davies, N. B. & Houston, A. (1984). Territory economics. In *Behavioural Ecology: An Evolutionary Approach* (2nd edn), ed. J. R. Krebs & N. B. Davies, pp. 148–69. Oxford: Blackwell Scientific.

Floody, O. R. (1983). Hormones and aggression in female mammals. In *Hormones and Aggressive Behavior*, ed. B. B. Svare, pp. 39–90. New York: Plenum.

Horn, H. S. & Rubenstein, D. I. (1983). Behavioural adaptations and life history. In *Behavioural Ecology: An Evolutionary Approach* (2nd edn), ed. J. R. Krebs & N. B. Davies, pp. 279–98. Oxford: Blackwell Scientific.

Hrdy, S. B. (1981). *The Woman That Never Evolved*. Cambridge, Massachusetts: Harvard University Press.

Trivers, R. L. (1972). Parental investment and sexual selection. In *Sexual Selection and the Descent of Man*, ed. B. Campbell, pp. 136–79. Chicago: Aldine.

8

Mechanisms of competitive aggression and their relation to function

In this chapter, we consider the mechanisms whereby competitive aggression achieves its function. The first section is concerned with general principles, in particular the design of a motivational system underlying territorial aggression, and the other forms competitive aggression may take (such as frustration-induced aggression). In section 8.2, we discuss the internal mechanisms which ensure that competitive aggression occurs only at functionally relevant times, notably when it is necessary to obtain resources for reproduction. In the remainder of the chapter, we discuss a wide range of mechanisms which may underlie the species and sex differences in aggression described in Chapter 7. Again, we consider these in relation to the categories outlined in Chapter 1, i.e. phylogenetic, developmental and short-term sources of variation.

8.1 General principles

The functional category 'competitive aggression' encompasses a wide range of mechanisms in different taxonomic groups. In Chapter 4, some general design principles were described, and these can be applied to competitive aggression. The simplest mechanism is one whose output would be triggered indiscriminately by a nearby chemical or tactile stimulus. This occurs in anthozoan coelenterates where the response continues in a stereotyped form until completion (Chapters 2 and 7).

Territorial aggression occurs in the large Californian limpet, *Lottia gigantia*, where the nervous system is also a diffuse nerve net (Borradaile *et al.*, 1961). Stimson (1970) showed that the limpet could distinguish between different classes of intruders, would continue the response (pushing movements) until the opponent had been dislodged from the rock, and would then cease. The mechanism is therefore initiated by recognition of specific stimuli, and its output continues until

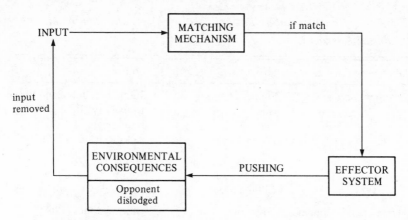

Fig. 8.1. Control system model of territorial aggression in the Californian limpet *Lottia gigantia*.

the initiating stimulus is no longer present. These design features can be represented by the simple feedback loop shown in Fig. 8.1 (and they contrast with the anthozoan example, where behaviour is not modified by its consequences).

In Chapter 4, I introduced a general control theory model of aggression (Fig. 4.2), which involved mismatch-detection on the input side, and attack, escape or both, as possible outputs, the consequences of which would be to remove the input (Archer, 1976*a*). This model was also applied to maternal aggression (Chapter 6). In a wide range of vertebrates, territorial aggression is likewise based on detecting a novel intruder into a familiar area (Fig. 8.2). Attack occurs only when the animal is in an area containing familiar environmental cues (e.g. Figler & Einhorn, 1983, for a cichlid; Kuo, 1960*b*, for quail; Dudzinski, Mykytowycz & Gambale, 1977, for rabbits). Attacks are directed to a wide range of novel intruders and novel objects (Table 8.1), and the rate of attack declines with repeated exposures to the same stimulus (e.g. Peeke & Peeke, 1970; Peeke, Herz & Gallagher, 1971; Peeke & Veno, 1973).

Wiepkema (1977) has put forward a similar model of territorial aggression. It involves a 'norm', or neural representation of the territory without the opponent, which is compared with the degree to which the opponent has entered the territory (referred to as 'provocation'). Essentially, the model consists of a negative feedback loop which involves comparing the input from the territory with the norm. The result of this comparison eventually activates a motor command for

Table 8.1. *Range of unfamiliar animals or objects attacked by an established resident*

Object of attack	Species	Source
Another species	Siamese fighting fish Cod *Peromyscus* Chimpanzee	Johnson & Johnson (1973) Brawn (1960) Baenninger (1973) van Lawick Goodall (1968)
Lame, deformed or paralysed conspecific	Uganda waterbuck Wild Norway rat Chimpanzee	Spinage (1969) Alberts & Galef (1973) van Lawick Goodall (1968)
Human hand or finger	Great tit Feral Norway rat Domestic fowl chick Housemouse	Blurton-Jones (1968) Galef (1970) Horridge (1970) Ebert (1983)
Moving cardboard box	Domestic fowl chick	Bateson (1964)
Moving cardboard rectangle	Several species of chicks	Schaller & Emlen (1962)
Bottle brush	Mice	Lagerspetz & Mettala (1965)
Pencil, paper roll	Great tit	Blurton-Jones (1968)
Green wooden triangle	Domestic fowl chicks	Evans (1967)
Red wooden model	Three-spined stickleback	Peeke, Wyers & Herz (1969)
Sudden noise	Cod	Brawn (1960)

aggressive behaviour. If the opponent is driven out, the value of the input decreases whereas if the opponent goes nearer to the centre of the territory, it increases, thus increasing the motor command. In addition to the main negative feedback loop, there is a positive feedback loop serving to keep up the momentum of the behaviour when the value of the input has momentarily declined, and a small negative feedback loop enabling the aggressive tendency to dissipate when no opponent is present.

Although there are some important differences in design between Wiepkema's model and my own general model (Fig. 4.2), both view aggression as a reaction to external events, based on a discrepancy-activated process, which has the effect of removing the initiating conditions. These models are therefore fundamentally different from

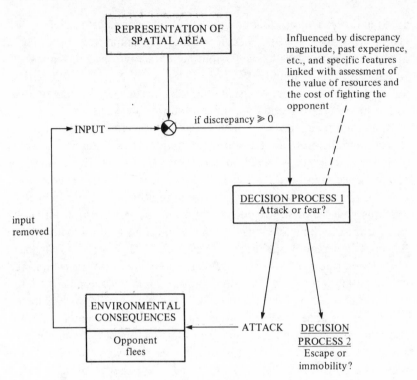

Fig. 8.2. Control system model of territorial aggression. (Based on Archer (1976*a*), Fig. 1.)

that of Lorenz (1950, 1966), who viewed aggression as an endogenous drive (Chapter 1).

Both models represent territorial aggression as a response to novel or unexpected events (i.e. involving a mismatch principle). In contrast, early ethological research emphasised the importance of specific stimuli ('social releasers') in initiating territorial attack (Lack, 1939; Tinbergen, 1951, 1965). More recent research has called into question Tinbergen's classic finding that territorial sticklebacks attack the red abdomen of an intruder (Peeke *et al.*, 1969; Rowland, 1982; Rowland & Sevenster, 1985). Chantry & Workman (1984) have also shown that the stimuli which evoke aggressive displays in the robin are more complex than Lack (1939) originally reported. In addition, territorial responses are now known to be complicated by status badges (e.g. Studd & Robertson, 1985; Hansen & Rohwer, 1986). Nevertheless, specific stimuli do appear to be important in some cases: for example, colour in lizards (Ferguson, 1966; Greenberg & Crews, 1983), fish (e.g. Kohda,

1983) and cephalopods (Tinbergen, 1965); electric fields in electric fish (Black-Cleworth, 1970). Specific odours are important in rodents, although specificity and novelty interact (Mackintosh & Grant, 1966; Mugford, 1973; Flannelly & Thor, 1978; R. J. & D. C. Blanchard, 1981). Another category of specific stimuli eliciting attack come to do so as a result of classical and operant conditioning (section 8.5).

It is therefore apparent that whatever the role of discrepancy-detection in the motivational control system for aggression, there are, in addition, mechanisms which involve matching with representations formed by innate biases or specific past experiences in at least some cases (Curio, 1975; Baerends, 1985). Where both novelty *and* a specific stimulus have been shown to be important, as in territorial attack by mice or rats, both matching and mismatch mechanisms would have to be depicted in a model of the control system. R. J. & D. C. Blanchard (1981) reported that when a stranger was placed into an established colony of laboratory rats, the dominant or alpha male approached and sniffed the intruder's perineal region. If the stranger was a subadult or had been castrated, it was not attacked. But if it was an adult male, the resident would show piloerection, and shortly afterwards would begin a sequence of attack, chase and bite. These observations suggest that there is a two-stage process in the decision whether to attack: first, a general investigation of the intruder, as any novel stimulus would be investigated, and secondly, a more specific investigation which must involve a matching process, since only one class of stimulus – the odour of a mature male – evokes piloerection and attack.

Competitive aggression may take other forms besides territorial attack. For example, an animal may attack another which comes too close (individual distance intrusion), or else compete directly for a resource. In some arthropods, such as crayfish and stick insects, the males attack and dislodge one another when copulating (Sivinski, 1978; Berrill & Arsenault, 1984). Similar behaviour is found in amphibians (Davies & Halliday, 1978; Smith & Ivins, 1986) and in rodents under crowded conditions (Louch, 1956).

One important motivational mechanism which underlies many examples of competitive aggression in birds and mammals is frustration (Chapter 1). When an animal expects a reinforcing event, anything that prevents access to such an event produces a state referred to as frustration (Amsel & Roussel, 1952). This concept had earlier been applied to human aggression, in the form of the frustration-aggression hypothesis (Chapter 1), which had a strong influence on later social psychological accounts of aggression (Berkowitz, 1962, 1969). From the

1960s onwards, studies were carried out to investigate the possible link between frustration and aggression in animals (reviewed by Archer, 1976a, Looney & Cohen, 1982). When a pigeon, rat or squirrel monkey is first trained on continuous reinforcement (CRF), and then placed on extinction, it will attack any suitable nearby target. Similar results are obtained if the CRF schedule is replaced by one involving a delay in obtaining reinforcement (such as fixed-ratio, fixed-interval or fixed-time), or if the learnt response is prevented by a physical barrier. The latter is similar to observations, made under field conditions, of birds or primates competing for localised packets of food (e.g. Hinde, 1953; Stokes, 1962; Chalmers, 1968; van Lawick Goodall, 1968). Yet food deprivation by itself does not increase aggressiveness, at least for a wide range of birds and mammals (Gottier, 1972; Archer, 1976a). This suggests that frustration, or the thwarting of an activated feeding tendency, is probably the mechanism underlying these examples of food competition.

Animals also fight over other resources, such as nesting materials (Crook & Butterfield, 1968; Robinson, 1986), or access to receptive mates (Arvola, Ilmen & Koponen, 1962). These examples may be mediated by frustration as well. However, experimental studies have involved few reinforcers other than food. Water (Gentry & Schaeffer, 1969; Campagnoni, Cohen & Yoburn, 1981), opiate drugs (Boska, Weisman & Thor, 1966; Davis & Khalsa, 1971) and electrical stimulation of the brain (Looney & Cohen, 1982) have been used, but not other known reinforcers, such as access to nest material, exploration, heat, a running wheel, or a sexual partner (Hogan & Roper, 1978). The generality of the link between frustration and aggression for all reinforcers has yet to be established (Roper, 1981). Similarly, the range of animals used in the experimental studies is restricted to a few species of birds and mammals.

Operant conditioning provides a further mechanism underlying competitive aggression in groups of animals. Aggressive responses which initially occurred to frustration or individual distance intrusion will be reinforced if they result in access to a resource such as food or water (Ulrich et al., 1963; Azrin & Hutchinson, 1967). This subject will be discussed further in section 8.5.

8.2 The internal control of competitive aggression

In Chapter 6, we noted that it was particularly important that there were mechanisms ensuring that parental aggressiveness occurred only at functionally appropriate times. These mechanisms took the form of either hormonal changes, or cues associated with the presence of the eggs or young, or both (section 6.3). Similar mechanisms are also necessary for competitive aggression. Frustration-induced aggression occurs only when the animal is pursuing a goal-directed activity, which usually involves obtaining a resource. Territorial behaviour is often dependent on the level of food resources in the territory (section 8.5).

In addition to such short-term fluctuations, which are associated with the importance of a resource to the animal and resource availability, there are also long-term changes in aggressiveness associated with reproductive condition. These changes are controlled by hormones, particularly those involved in sexual maturation, and changes throughout the breeding cycle.

Although there are few studies of endocrine influences on behaviour in invertebrates, it has been shown that juvenile hormone levels and ecdysteroids from the ovary influence dominance in females of the primitively social wasp *Polistes gallicus* at the beginning of the annual cycle (Roseler *et al.*, 1984, 1986).

The influence of testosterone and other androgens on aggression in male vertebrates has been studied extensively (Brain, 1977, 1979, 1981*a*). In general, androgens increase aggressiveness, although this may be overridden by habit or experience, and the precise influence varies widely between different species (section 8.3), and the sexes (Archer, 1976*a*; Floody, 1983). Androgen levels generally increase at times when sexual competition or competition to obtain resources for reproduction becomes functionally appropriate. This represents an indirect relationship between function and causal mechanism based on a temporal coincidence of mechanism and result (Chapter 1).

Androgens such as testosterone may affect aggressiveness in one of several ways. The first possibility is a direct effect on the decision whether to attack or to flee, making attack more likely. Testosterone does not appear to increase or decrease fear responding (Archer, 1975, 1976*b*; Beatty, 1979), or affect submissive behaviour (Schuurman, 1980). Figure 4.5 (Chapter 4) shows the influence of a variable – such as testosterone – which increases the tendency to attack over a wide range of inputs, but does not affect fear: this results in the point where fear overtakes attack being moved to a higher input level, and also to a lowered attack threshold.

The other influences of testosterone on aggression are more indirect. The first concerns territory formation. As indicated in section 7.5, territory formation coincides developmentally with a rise in androgen levels in many *K*-selected species (Brain, 1977, 1979). In species which show a seasonal transition from non-territorial to territorial behaviour, a rise in androgen levels is associated with the transition (Brain, 1977, 1979; Harding, 1983; Villars, 1983). Marler (1956*a*, *b*), Mackintosh (1970), and Walther (1977) have described the enlargement of individual distance to form a territory that occurs at the beginning of the breeding season for chaffinches, mice and Thompson's gazelle, respectively. In these cases, some initial internal change must have produced an expansion of the animal's individual distance until it became linked to an object or objects in the environment. As Hinde (1970) commented, the change in internal state 'must influence the relevance of stimuli rather than the intensity of behaviour'. Stimuli which previously had little significance now provide landmarks and boundaries (De Boer & Heuts, 1973; Mackintosh, 1973; Figler *et al.*, 1975), and these in turn influence whether an aggressive response is shown. Such stimuli appear to be used to specify a desired state, such as 'no birds near my trees' (Hinde, 1970, p. 342), which corresponds to 'the norm' in Wiepkema's model (section 8.1). How testosterone facilitates such an internal change is at present unknown.

In experimental studies of chicks, mice and rats, testosterone has been shown to enhance attention towards stimuli which are controlling the current behaviour and to reduce attention to other stimuli (Andrew, 1972; Andrew & Rogers, 1972; Archer, 1974, 1977*b*; Thompson & Wright, 1979). More recent studies have shown that these changes involve effects on memory formation which are also produced by other gonadal steroids (Andrew, 1979, 1983; Clifton, Andrew & Rainey, 1986). A period of sustained attention to the opponent occurs prior to attack in many species, and it is possible that testosterone facilitates this by reducing distractability (Thor, 1980). The necessity for such a period of sustained attention may be related to the observation that a period of sensitisation or 'warming up' often precedes full attack (Peeke & Veno, 1976; Peeke, 1982; Potegal & tenBrink, 1984). Clifton *et al.* (1986) have shown that the effects of testosterone on attack, and on attentional and memory processes, occur at different latencies in the domestic chick, and are likely to be dependent on different mechanisms.

A further indirect effect of testosterone on aggression may occur through its anabolic influence on muscular development. Kuo (1960*a*) found that quail given protein-enriched diets showed increased readiness to attack. Bevan, Davies & Levy (1960) found that the extent to which aggressiveness declined following castration in male mice was positively related to their decline in body weight. Since larger animals are more likely both to initiate and to win fights (Chapter 9), any effect of testosterone mediated through changes in body weight could be related to this effect of size, which operates through the responses of other animals.

Other indirect influences of testosterone may also be mediated by changes in social presentation, since there are many androgen-dependent peripheral receptor sites, which are important for social behaviour. These include mammalian scent glands (Mykytowycz, 1962, 1966; Mugford & Nowell, 1971), avian vocal structures (Andrew, 1969), the combs of Galliformes (Moss *et al.*, 1979; Harding, 1983), and the antlers of red deer (Lincoln *et al.*, 1972). The importance of such social cues is illustrated by a study by Marler (1956*a*), who disguised female chaffinches as males by dyeing their breast feathers red. He found that they were treated as males by other birds and they soon came to behave like males, attacking at greater distances than was the case for normal females. Rohwer & Rohwer (1978) dyed the plumage of subordinate white-crowned sparrows so that they looked like dominant birds. In this case, they were simply attacked more and did not become dominant. If, however, they were injected with testosterone as well as having dyed plumage, they were treated as dominant by the existing dominant birds. Rohwer & Rohwer explained their results in terms of the birds' behaviour and disposition having to match their external appearance in order for successful 'deception' to occur. However, in a later study on Harris' sparrows (Rohwer, 1985), it was found that when young birds were dyed to resemble dominant adults, they did achieve domination over their age-mates, by a two-stage process: the first involved the dyed birds being avoided by the other young birds, and the second, shortly afterwards, involved active displacement.

Most of the evidence on the relationship between hormones and aggression in females is restricted to mammals, and there are wide species differences, unlike the general facilitating influence of testosterone in the male. These species differences will be described in the next section.

Pituitary–adrenocortical influences on aggression in mice have been investigated extensively since the early 1970s (Chapter 1). There appear to be two main effects associated with the activity of this system, one long-term and the other short-term (Leshner, 1983*a*). The long-term influence is to decrease aggressivensss between male mice, and this appears to be a direct result of higher levels of ACTH on behaviour (rather than an indirect consequence of changes in the level of another hormone). Its behavioural consequences seem to involve a specific decrease in aggressiveness, rather than an indirect effect through changes in fear or submission. This corresponds to a change from A_2 to A_1 in Fig. 4.5, which is the opposite to the one which was suggested above for the influence of testosterone.

There are, however, some reservations about the findings for ACTH (Leshner, 1983*a*). The effect is not a very robust one, it is sensitive to strain differences and changes in the testing procedure. One study also found that *low* levels of ACTH appear to be associated with decreased aggressiveness as well, raising the possibility of a biphasic effect.

It is not clear in functional terms why high levels of ACTH should decrease aggressiveness in mice. The most obvious possibility is that ACTH levels are higher in times of stress and a decrease in aggressiveness will avoid further damage and defeat.

The short-term influence of pituitary–adrenocortical hormones will be discussed in section 8.5 (see also section 4.7).

8.3 Mechanisms underlying interspecific and sex differences in responsiveness

In the next three sections, we discuss the way differences in general design features and internal control mechanisms can account for some of the species, sex or age differences discussed in Chapter 7. In this section, we are concerned with mechanisms whose flexibility lies at the phylogenetic level, again following the distinction between adaptive mechanisms made in Chapter 1. Mechanisms involving developmental and short-term flexibility will be discussed in sections 8.4 and 8.5.

Most of the experimental evidence showing that individual differences in aggressive behaviour are heritable comes from selective breeding experiments carried out on rodents. Earlier studies in the comparative psychological tradition had shown marked strain differences in the aggressive behaviour of laboratory rats and mice (Lagerspetz, 1964; McClearn & DeFries, 1973). Lagerspetz (1964) selectively bred male mice for high and low levels of aggression, and Lagerspetz & Wuorinen (1965) showed that mice from the aggressive strain would maintain their characteristic after being cross-fostered to mothers from the non-aggressive strain. Ebert & Hyde (1976) initiated a similar selective breeding programme using wild females, and again found that aggressiveness could be selected artificially (Ebert, 1983). Bakker (1985) has obtained similar results for male and female sticklebacks.

These limited studies show that aggressiveness can be subject to artificial selection. Many of the interspecific differences considered in this section have presumably resulted from parallel selective processes operating under natural conditions.

Differences in aggressive behaviour between species and the sexes can take many forms, such as whether it occurs at all, when it is likely to occur, the range of eliciting stimuli, the latency of responding and the strength and form of the response. The possible sources of such differences include the following:

1 alteration of the relationship between the input and the relative motivational strengths of aggression and fear;
2 the nature of the comparison process which monitors the input;
3 the type of internal representations used in this comparison process;
4 internal influences resulting from hormonal changes and past experience;
5 the properties of the opponent or target (which will be discussed in Chapters 9 and 10).

In Chapters 4 and 6, I suggested that the level of input at which the tendency to attack was replaced by that for fear was an important variable in accounting for interspecific and intersexual differences (see Figs 4.3 to 4.9). In particular, it was argued that selection for a decrease in fear tendency (i.e. an increase in 'boldness') could indirectly increase aggressiveness by raising the threshold for fear behaviour. Similar principles apply to competitive aggression.

A second possible source of interspecific differences is whether the comparison process involves a mismatch or matching mechanism (section 8.1). In many animals, the territory-holder investigates the intruder for a period of time prior to attack (as described for rats in section 8.1). Studying two species of vole, de Jonge (1983) found that the more solitary one, *Microtus agrestis*, showed little olfactory investigation before attacking a conspecific, whereas the more social one, *M. arvalis*, would spend a longer time investigating the opponent. Beilharz & Beilharz (1975) also found longer attack latencies in laboratory-bred mice than in wild-caught mice. It is therefore possible that in solitary species the decision process leading to attack depends largely on a mismatch mechanism, whereas in social species a complex decision process, consisting of both mismatch and matching to specific stimuli, is involved.

A third possibility concerns the representations used to assess the input. When a fish or rodent establishes its territory, certain features of its environment become represented internally, and (presumably) are used as reference points when assessing whether a territorial intrusion has occurred (section 8.3). There are clear differences between species of fish in the degree to which familiar landmarks confer an advantage in an encounter with a strange conspecific. This occurs more readily in the jewel fish (Figler, Klein & Peeke, 1976) than in convict cichlids (Figler & Einhorn, 1983). In these species, the time-scale is a matter of hours, but in two field studies, of speckled butterflies and red-winged blackbirds, reversal of territorial ownership has been observed to take place within minutes (Davies, 1978*b*; Hansen & Rohwer, 1986).

It would appear, therefore, that some species more readily 'lock on' to landmarks, i.e. they become rapidly incorporated into the animal's internal representations. The possible role of testosterone in facilitating this process was discussed in section 8.2 (cf. Bakker, 1986, who found evidence that selection for territorial aggressiveness in sticklebacks may operate on androgen levels).

A variety of internal influences on aggressiveness, such as hormones (section 8.2), may vary in different species and in the two sexes. In particular, differences in the timing of androgen secretion, their levels, and their influence on, and the presence of, receptors can account for much of the variability in aggression between different species, sexes and ages in vertebrates.

Although a facilitative link between androgens and aggressiveness appears to be widespread in vertebrates (section 8.2), there are some exceptions. Among teleosts, it may be restricted to certain groups, such as cichlids (e.g. Munro & Pitcher, 1985), but absent from others such as centrarchids and Gasterosteiformes (Villars, 1983). In the centrarchids, androgens seem to influence site establishment and nest building but do not exert a direct effect on aggressiveness (Villars, 1983).

Among amphibians and reptiles, the subject of androgens and aggression is largely unexplored, with the exception of studies of the lizard *Anolis carolinensis*, which have produced conflicting results (Greenberg & Crews, 1983).

In birds, there are far more studies, and Harding (1983) concluded that testosterone is the most important hormone affecting aggression in this group. There is evidence that testosterone can increase aggressiveness in the male for the following birds: mallards, domestic fowl, red grouse, starlings, quail, redwings, black-crowned night herons, white-crowned sparrows, pigeons and weaver birds. The avian ovary also secretes high levels of androgens: female territorial behaviour has been associated with androgens in a variety of species, including both those where there is reverse sexual dimorphism, such as phalaropes (section 7.5), and those where there is not, such as robins and white-crowned sparrows (Harding, 1983).

In mammals, the facilitating influence of testosterone on aggression is well known but studies have concentrated on convenient or economically important groups such as rodents (Brain, 1977, 1979, 1981a) and ungulates (Bouissou, 1983), and there are relatively few studies on other groups, including primates (Bernstein et al., 1983).

It has been suggested that gonadotrophic hormones might facilitate competitive aggression in some fish, principally sticklebacks (Hoar, 1962; Baggerman, 1966; Wootton, 1970), and in birds such as starlings and weaver birds where LH has been implicated (Davis, 1957; Matthewson, 1961; Crook & Butterfield, 1968; Lazarus & Crook, 1973). However, as Villars (1983) pointed out, the evidence for teleosts is still limited and circumstantial. There are a number of methodological problems involved in interpreting studies of hormonal influences in fish. For example, if castration has no effect on aggressive behaviour, this does not necessarily mean that another hormone, such as a gonadotrophin, is involved, since testicular regeneration occurs readily in many species. Also, pharmacological manipulations which inhibit gonadotrophins may exert a direct effect on aggressive behaviour. The evidence that LH facilitates aggression in starlings and weaver birds is restricted to the four studies cited above, and is all indirect. Again, there are methodological problems, and Harding (1983) concluded that the LH hypothesis is unlikely to be correct. As indicated above, more recent evidence suggests that testosterone facilitates attack in passerines, including starlings.

In the previous section we noted that there are interspecific differences in the nature of hormonal influences on aggression in female mammals. Edwards (1969) and Gray, Whitsett & Ziesensis (1978) found that testosterone facilitates aggression directed towards juvenile intruders by group-housed female housemice. It seems that this androgenic influence is specific to attack of juveniles, which also occurs in females of other rodent species (Ayer & Whitsett, 1980; Floody, 1983).

More generally, it is found that progesterone, or a combination of progesterone and oestrogen, facilitates aggression in female rodents. In the golden hamster, a dioestrous female tends to win encounters with a male (Payne & Swanson, 1970), and there is some evidence that progesterone facilitates this (Payne & Swanson, 1971, 1972; Takahashi & Lisk, 1983). Neither oestrogen nor testosterone alone will affect the aggressiveness of ovariectomised hamsters (Payne & Swanson, 1971, 1972; Vandenberg, 1971), but a combination of oestrogen and progesterone lowers aggressiveness (Floody, 1983), thus providing a mechanism whereby aggressivensss towards males is lowered in oestrous females (Chapter 7).

In the brown lemming, there is a dramatic facilitating effect of progesterone treatment, and Floody (1983) speculated that this could be part of a mechanism to terminate oestrus in response to stimulation received during mating. In the white-footed mouse (*Peromyscus leucopus*), Gleason, Michael & Christian (1979) found that ovariectomy lowered aggressiveness and increased submissive behaviour. High levels of either testosterone or progesterone (but not oestrogen) increased aggressiveness, but neither hormone affected submissiveness.

In contrast to these findings which suggest a facilitative effect of progesterone, Barfield (1984) suggested that the reduced aggression towards colony intruders observed at oestrus in laboratory rats (Hood, 1984) was probably a consequence of increased progesterone, and the subsequent increase in aggressiveness at dioestrus was due to a decline in progesterone.

Few definite conclusions emerge from the existing evidence on hormones in female rodents, except that progesterone and testosterone affect aggressiveness in some cases. Female hormonal influences appear more varied across species than those of the male, where testosterone facilitates aggression from teleosts to mammals. In future research on females, it would be useful if the various hormonal mechanisms could be related to the function of aggression at different stages of the animal's reproductive life.

Until recently, the mouse was the only species in which the long-term influence of pituitary–adrenocortical hormones had been studied. Munro & Pitcher (1985) have now studied the influence of cortisol in a cichlid, but their results were not entirely clear-cut.

This section has been concerned with mechanisms which might underlie species and sex differences in territorial and related forms of aggression. For frustration-induced aggression, there is little evidence for species differences (Looney & Cohen, 1982). Frustration-induced aggression is also readily obtained from female as well as male mice and rats (Fredericson & Birnbaum, 1954; Looney & Cohen, 1982). Hahn (1983) did, however, report large strain differences in mice for latencies to attack in food competition test (which is equivalent to frustration-induced fighting: Adams, 1979).

8.4 Developmental flexibility

There is no strong tradition of research into developmental influences on animal aggression, and there is little overall framework to hold the scattered evidence together. Commenting on this, Bekoff (1981) advocated longitudinal behavioural studies carried out from ecological and evolutionary perspectives, and he reviewed studies which showed that the social development of bighorn sheep (*Ovis canadensis*) varied in different habitats, that food resources can affect the development of aggressive behaviour in the medaka fish (*Oryzias latipes*) and in ospreys (*Pandion halietus*). In addition, a wider variety of ecological conditions, including predation pressure, can affect the development of aggressive behaviour in cockroaches.

In contrast to these few ecologically based studies, there is a variety of laboratory studies on rodents which demonstrate striking changes in their aggressive behaviour following specific prenatal and postnatal treatments. Perhaps the most important and comprehensive of these are the experiments of vom Saal on intrauterine position, which were considered briefly in Chapter 6. vom Saal (1984) suggested that gonadal hormones, which are passed from one foetal blood system to the next in mice, result in the development of phenotypic variation which is adapted to either low or high population density. His theory is a modification of an earlier one (Chitty, 1960), put forward to explain rodent population cycles. The original theory involved genetic polymorphism, but vom Saal suggested developmental flexibility instead. Given the time span of the changes, this appears more likely, particularly as a plausible mechanism has been demonstrated.

In Chapter 6, I described vom Saal's finding that female mice whose intrauterine position was between two males became more maternally aggressive (but less fertile) than those which were located between a male and a female or between two females. These differences in aggressiveness extend to encounters between females which are not pregnant or lactating. vom Saal argued that the more aggressive females, which would occur with an average frequency of 1 in 6, would be better adapted for high-density populations where competition for space to breed would be more important than fecundity. However, at low densities, the more common, less aggressive but more fecund females would have the advantage. vom Saal suggested that a mechanism for the transport of foetal gonadal steroids had evolved specifically to produce this developmental polymorphism in microtine rodents, thus enabling some of the offspring to be adapted to dense populations but the majority to lower densities. (In order to substantiate this functional hypothesis, it would be necessary to demonstrate that such animals did leave more descendants than mice producing only one or the other type of offspring.)

Prenatal stress caused females which were not positioned between two males to become more aggressive, a possible mechanism being a rise in foetal testosterone levels which occurred in stressed animals of both sexes around day 17 of gestation. vom Saal suggested that this provides a further means of producing more aggressive but less fertile females better adapted for high densities. He also demonstrated that a male which developed between two males, and was gonadectomized at birth, would attack more readily when treated with testosterone in adulthood than a similarly treated animal from another intrauterine position. When intact, such males show less sexual mounting and fewer infanticidal tendencies than males which developed between two females. Prenatal stress results in males which developed between two females coming to resemble those which had developed between two males in the absence of stress. Again, vom Saal suggested that the function of these changes was to enable males to adapt to the different densities found in populations of microtine rodents. We should, however, be cautious of the prenatal stress findings, since Harvey & Chevins (1985) and Kinsley & Svare (1986) have reported effects on aggressiveness which are apparently in the opposite direction, i.e. that male mice were less aggressive following stress. In earlier research, the direction of the influence of prenatal stress on emotional reactivity was variable, depending on factors such as genetic make-up and early experience (Archer & Blackman, 1971).

In motivational terms, intrauterine influences represent long-term changes in responsiveness, perhaps involving a variable such as the relative strengths of aggression and fear tendencies, or the degree to which the animal becomes attached to a spatial area.

There are many studies of the influence of restricted social environments during rearing. The subject has generated considerable interest in relation to two issues: first, the attempt to separate environmental from hereditary influences on behaviour; secondly, its possible relevance to the developmental origins of human psychological disorders (Bowlby, 1969, 1973).

Early postnatal influences on adult aggressiveness were studied in a number of older comparative psychological studies. Some have shown that prolonged rearing in social isolation often results in increased aggressiveness, in fish (Kuo, 1967), quail (Kuo, 1960c), Asian song thrushes, dogs (Kuo, 1967) and rhesus monkeys (Mason, 1960, 1963; Harlow *et al.*, 1965), although others show increased fearfulness instead (Harlow *et al.*, 1965; Fuller, 1967; Kuo, 1967).

Most other studies of early influences have been carried out on rodents where, until recently, the presence of the mother was essential for survival. Consequently, complete isolation was not practical before weaning (cf. Day *et al.*, 1982). A number of studies have found that adult aggression is increased in male mice following an absence of peer-group contact between birth and weaning (Denenberg, 1973). These results suggest that social interactions between male littermates exert an inhibiting effect on their later aggressive behaviour. Comparing males which had been reared with male or female littermates, Namikas & Wehmer (1978) found that those reared with females were more aggressive in young adulthood, again suggesting an inhibiting effect of social interactions between male littermates. Presumably, they learn to restrain their attacks on other males, as a result of punishment (Taylor, 1979), and the development of fighting strategies (Bekoff, 1981; Chapter 10). In this context, it is interesting to note that the heightened aggressiveness found in socially isolated rhesus monkeys may show itself in the form of inappropriate attack which ignore signals of age and size (Bernstein, 1981).

Several studies have investigated the effects of cross-fostering mice or rats to a female of an inbred strain differing in aggressiveness. When male mice from a non-aggressive strain were fostered to mothers from an aggressive strain, their adult aggressiveness was greater than that of mice fostered to their own strain (Southwick, 1967, 1968). Denenberg (1973) cross-fostered mice to rat mothers and their offspring and found a strain-dependent influence on adult aggression. *C57B1* mice, which are normally aggressive, showed very little adult aggression after being reared by rats, whereas Swiss albinos, which are normally less aggressive, showed little influence of being reared by rats. These influences were neither due to social interaction with rat pups nor to the effects of fostering *per se*, but seem to be connected with a preference for rats and a lack of interest in mice by the fostered animals (Lore & Takahashi, 1984).

The grasshopper mouse (*Onychomys torridus*) is normally aggressive, but when reared by females of the more passive white-footed mouse (*Peromyscus leucopus*), it becomes less aggressive (McCarty & Southwick, 1979). Cross-fostering in the opposite direction (white-footed mouse to grasshopper mouse mothers) produced little change in this less aggressive species.

The influence of the father during the preweaning period has also been investigated. Male mice reared with both parents become more aggressive than those reared only by their mothers (Mugford & Nowell, 1972; Wuensch & Cooper, 1981). The increase in aggression occurred even in the absence of tactile contact with the father (Wuensch & Cooper, 1981), raising the possibility of an olfactory imprinting-like process. The researchers suggested that experience with an adult male during the preweaning period establishes a standard, and that during adulthood, discrepancies from this standard evoke aggression (cf. the findings of Bateson, 1978, 1980, 1982, on mate-choice in quail).

Rearing mice in a restricted social environment after they have been weaned may have a different influence on their subsequent aggressiveness to that produced by preweaning isolation from peers (Denenberg, 1983; see above). It produces a decrease in aggression in males (King & Gurney, 1954; King, 1957; Denenberg, 1973), which can be reversed by social stimulation such as 10 days' housing with male siblings, or 25 days' non-tactile contact with both male siblings and the father (King & Gurney, 1957), or non-tactile contact with strange adults (de Ghett, 1975). de Ghett also found that if these otherwise isolated males observed the strange adults fighting, a further increase in aggression occurred when they were tested as young adults.

It seems from these studies that isolation immediately after weaning leads to low levels of adult aggressiveness, and that some social stimulation is necessary during this period in order to produce the higher, more usual, level of adult aggressiveness.

In contrast to these studies, a number of others have shown that isolation-housing can *increase* aggressiveness in laboratory rodents (e.g. Yen, Day & Sigg, 1962; Hatch *et al.*, 1963; Sigg, Day & Colombo, 1966; Goldsmith, Brain & Benton, 1976; Brain & Benton, 1983). These conflicting findings are reminiscent of the effects found in the older comparative psychological studies (see above), where isolation produced either fear or attack. Indeed, some studies have shown that defence or flight are common alternatives to attack in postnatally isolated mice (Krsiak & Borgesova, 1973).

The results of a recent study by Cairns *et al.* (1985) may reconcile these conflicting findings. They investigated the effect of different durations of isolation on male mice, beginning at different ages. Although, in general, isolation-reared mice attacked a test animal more than group-reared ones, the earliest age of isolation onset (21 days) combined with the longest duration produced mice which were too reactive or immobile to attack effectively. Cairns *et al.* also found that the younger the age at which isolation had begun, the more powerful were its effects on measures of emotional reactivity such as startle and freezing. These results can be interpreted in terms of the general model of aggression introduced earlier (Chapter 4), as follows. Mice reared in isolation from an early age or for a long time develop very restricted representations of the environment. When exposed to a group-housed mouse in a strange environment (as occurred in the subsequent aggression test), the discrepancy between current input and these representations is very great, so that the input would exceed the point at which fear replaces attack (Figs 4.3 and 4.4). When isolation begins later or lasts for a shorter length of time, the representations will not be so restricted, so that the point at which fear replaces attack will be less likely to be reached.

In those studies finding lower levels of aggressiveness in mice isolated after weaning (e.g. King & Gurney, 1957), it may be that the animals behaved in a way similar to the early isolates in Cairns *et al.*'s study, i.e. they became highly fearful. The various forms of social stimulation used in these studies would have had the effect of widening the otherwise restricted representations of the environment.

There are few studies which seek to investigate developmental influences other than social restrictions during rearing. One of these is by Lore & Stipo-Flaherty (1984), who compared the aggressive behaviour of male rats housed in constant groups of three with those whose group membership was randomly changed at weekly intervals (for 10 weeks). After 2 weeks of isolation, the rats were tested with an unfamiliar conspecific. Only one of the three from the stable groups attacked the intruder, but all three from the random groups did so. A high degree of individual consistency was found when the rats were retested at 9 months of age. This study suggests that a pattern of dominance–subordination resulted from a consistent pattern of social interactions (which was possible only in the stable groups), rather than being a property of the individuals (Bernstein, 1981; Chalmers, 1981). Dominance relations were also developmentally stable, which is consistent with other reports in social mammals (Geist, 1978), including humans (Olweus, 1979, 1984).

In *K*-selected species, the onset of the adult pattern of competitive aggression usually coincides with sexual maturation (Chapter 7). In many mammals, the timing of this is subject to social influences. One example comes from a study by Epple (1981) on the saddle-backed tamarin, a monogamous New World monkey. Tamarins housed with an adult of the opposite sex from 6 months of age, for 4 to 6 months, showed earlier sexual maturity. They came to act like experienced pair-bonded adults, attacking strangers, and scent-marking. Young tamarins housed with a same-sex adult did not develop adult behaviour.

The experimental studies reviewed in this section are largely concerned with experimental restriction in laboratory mice and rats. They allow us only a few glimpses of the complexity of the interactions between a variety of social influences at different stages of development. It is clear that Bekoff's (1981) appeal for a more ethologically oriented approach to development was justified. More generally, a neglect of developmental flexibility in the study of aggression may have led to some rather rigid views of adult aggression, with an undue emphasis on either species-specific reactions or immediate stimuli (Adams, 1980).

8.5 Short-term mechanisms

In this section, we first consider the influence of food availability, distribution and levels on aggressive behaviour, and discuss the learning mechanisms underlying changes in behaviour occurring in response to food availability. We then briefly discuss the mechanisms producing dominance relations, and consider the question of whether aggression is reinforcing. Finally, short-term changes in aggressiveness mediated by hormones are outlined.

The issue of how flexible animals are in responding to short-term changes in the costs and benefits of their behaviour was introduced in Chapter 7 in relation to food competition. One fairly obvious form of flexibility is to abandon fighting when food is either widely distributed or superabundant. In the first case, it is achieved because animals become widely spaced as a consequence of spending most of their time foraging for food (e.g. Mykytowycz, 1961). In the second, it is achieved by removal of the conditions necessary for frustration, or by abandoning territoriality, which occurs during food superabundance even in species where it is normally a stable pattern of behaviour (Archer, 1970: the housemouse; Davies & Houston, 1984: the pied wagtail). The occurrence of territorial behaviour shows a closer parallel with food levels in many species. Wilcox & Ruckdeschel (1982), for example, found that adult and nymph water striders would cease being territorial when fed, and previously satiated animals which were then food-deprived began showing territorial behaviour 12 hours to 2 days later. Similar adjustments of territorial behaviour in relation to changing levels of food resources were shown in the studies of nectar-feeding birds described in Chapter 7.

Other studies have involved manipulating the spatial distribution of food. Zahavi (1971) found a change from flocking to territorial behaviour in overwintering white wagtails when he provided piles of food rather than a continuous spatial distribution. Similarly, Rubenstein (1981) found a greater tendency to defend territories in male pygmy sunfish when food was centrally rather than randomly distributed. Craig & Douglas (1986), however, found dominance relations at a concentrated nectar source but individual territories when the food was more dispersed in the bellbird.

In seeking to understand the mechanisms underlying the change from territorial to non-territorial behaviour as food resources change, it is convenient to consider Gill & Wolf's (1975) study of nectar-feeding birds. They found that the territory size of the Kenyan golden-winged sunbird was regulated so as to provide a relatively constant number of flowers. Defence of an area containing flowers raised its food availability by reducing the number of feeds taken by other birds. The nectar content of flowers within a territory was found to be two to four times greater than those outside, thus considerably reducing the time spent and energy expended in foraging. Gill & Wolf calculated that when the nectar levels of undefended flowers were low, even a slight increase in nectar availability from the defended flowers would justify the energy expended in territorial defence. When the nectar levels of undefended flowers were high, a substantial (and, in practice, unattainable) increase in nectar availability would be required from the defended flowers to make territorial defence economic. It was, therefore, predicted that the birds would be territorial only at lower nectar levels, and this was indeed found to be the case. Gill & Wolf also observed that when defence costs became high owing to a larger number of birds invading the territory, territorial behaviour again ceased.

These adaptive changes can be understood in terms of principles derived from studies of animal learning. It is known that attack can be subject to the influence of conventional reinforcers such as food or water. Hungry or thirsty pigeons or rats will increase their rates of attack on a conspecific or an inanimate target when provided with food or water reinforcement (Reynolds *et al.*, 1963; Ulrich *et al.*, 1963; Azrin & Hutchinson, 1967).

Using the more conventional operant response of bar-pressing, rats have been shown to increase responding on high-ratio schedules, so as to maintain the same food intake (Collier, Hirsh & Hamlin, 1972) provided there is no choice of another schedule. Similarly, if the size of reinforcement is small at any one time, rats will adjust to this by increased response rates (Collier *et al.*, 1972). Even so, very high schedule requirements do lead to abandonment of the schedule, despite there being no alternative (Lea, 1978).

In cases where two alternative interval schedules are available side by side, pigeons tend to match their responses to the rate or amount of food provided by the schedules, i.e. they respond to both schedules intermittently but in proportion to the amount of reinforcement obtained (Fantino & Logan, 1979). However, where reinforcements become available as a result of the animal's own actions rather than the passage of time (as in concurrent ratio schedules), the animal concentrates on the schedule leading to the greater magnitude of reinforcement.

The conditions under which Gill & Wolf's sunbirds were living can be viewed as a choice between two interval schedules, to feed from either defended or undefended flowers. Food becomes available more rapidly from defended flowers, and hence their reinforcement rate is higher. If the two types of flower were side by side, and undefended, the birds would match their responses to the different reinforcement rates. There are, however, two further considerations: first, the cost of switching between two spatially separated schedules, and secondly, the cost of defence. The first is straightforward in that the reluctance to change increases as the schedules become further apart (Larkin & McFarland, 1978). The second is more complex. Territorial aggression can be viewed as an operant response which provides access to the flowers with the higher reinforcement rate. As the number of responses necessary for access increases, the bird will become more likely to abandon this 'high-ratio' schedule in favour of the freely available but less productive undefended flowers.

As indicated in Chapter 7, the flexibility shown by nectar-feeders such as the sunbird contrasts with the behaviour of birds such as overwintering pied wagtails and ural owls, which show stable territorial behaviour despite changes in food availability. Here, territorial defence does not represent a choice between different schedules of reinforcement, but is constrained by a high degree of attachment to the spatial area (section 8.2). Presumably, a strong site attachment, like a strong social attachment, involves an affective component, and controls and supersedes responsiveness to immediate changes in 'economic' conditions.

Where animals are organised into groups or flocks, and their food is unevenly distributed but not superabundant, aggressive behaviour is likely to be mediated by frustration (section 8.2) and by instrumental conditioning. Where the group is stable, and individual recognition is possible, dominance relations develop as a result of the outcome of fights, by the processes of operant and classical conditioning. Bernstein (1981) has argued that dominance results from the specific experience of being defeated by, or defeating, a particular individual. Others have argued that animals come to undergo a general modification of responsiveness following a consistent history of victories or defeats and that discriminatory stimuli are simply more specific when learning dominance relations (Flannelly & Blanchard, 1981).

Another issue raised by the modification of aggressive behaviour following the outcome of a fight is whether winning is reinforcing when it does not provide access to conventional reinforcers such as food. Many of the studies investigating the reinforcing properties of aggressive behaviour have used Siamese fighting fish, *Betta splendens*, swimming through a ring for the opportunity to perform an aggressive display to a mirror image as the reinforcer (Hogan, 1967; Hogan & Roper, 1978). Johnson & Johnson (1973) showed that *Betta* would perform this response to viewing not only a mirror-image and another male, but also other objects such as a marble and even an empty chamber. They suggested that visual exploration, rather than the opportunity to perform the display, had provided the reinforcement in the earlier studies. However, Bols (1977) found that although several stimuli supported swimming in a T-maze, the level of performance was positively related to the extent to which an aggressive display was evoked, and Bols concluded that aggressive motivation, rather than opportunity for visual exploration, was the main motivating factor.

Many older studies on mice also indicate that winning a fight increases subsequent readiness to attack (Ginsburg & Allee, 1942; Scott, 1946; Lagerspetz, 1964). Experiments by Tellegen, Horn & Legrand (1969), Legrand (1970) and Tellegen & Horn (1972) have shown that the extent to which the opportunity to attack a submissive mouse provides reinforcement is dependent on the subject's level of aggressive motivation, suggesting that this is the crucial determinant of the level of performance.

There are two views concerning the nature of aggressive reinforcement, that it is provided either by the opponent's escape, or by the opportunity to attack in the absence of punishment. The first, shown in Table 8.2, is essentially a negative reinforcement view. Aggression is considered as a reaction to an aversive state, such as frustration (in the case of offensive attack), or pain (in the case of defensive attack). Removal of the aversive state produces a cessation of attack. This view

Table 8.2. *Negative reinforcement view of aggressive behaviour*

		LEVEL OF AVERSION	
		HIGH	LOW
INSTRUMENTAL RESPONSE	ATTACK	Fear-induced ('defensive') attack	Anger-induced ('offensive') attack
	ESCAPE	High-intensity escape and fear	Low-intensity avoidance

can be extended to cover cases where the opponent is not physically removed, but shows submissive postures instead. For example, in a study of *Betta*, Melvin (1985) found that the hurdle-crossing response was no longer maintained either when the target fish was absent or when it showed a submissive response.

The negative reinforcement view does not, however, explain why an animal will perform an operant response to obtain the opportunity to attack an opponent. Consequently, offensive forms of attack have often been viewed as appetitive (Hinde, 1970; Rasa, 1976), i.e. positively reinforcing. In studies of mice which learned a T-maze for the opportunity to attack a submissive opponent (Tellegen *et al.*, 1969; Legrand, 1970; Tellegen & Horn, 1972), maze-running was greatly facilitated in mice primed by being given a pretest fight and in those rated as 'highly aggressive', or from an aggressive strain. An interesting parallel to these results is found in human laboratory studies of aggression: Edmunds & Kendrick (1980) found that the apparent pain of a victim served to reinforce aggressive responses only in those subjects who had scored highly on questionnaire measures of aggression and hostility. Feshbach *et al.* (1967) found that only when subjects were angered did the apparent pain cues from the instigator of the anger act as a reinforcer for aggressive responses. These findings all suggest that the motivational or temperamental state of the subject is crucially important for establishing positively reinforcing effects of aggression. Potegal (1979) has also argued that aggression becomes substantially reinforcing only when an internal motivational state has reached a high level as a result of sensitisation or priming.

The two views of aggressive reinforcement outlined above have different implications for a motivational model of aggression. Negative reinforcement is consistent with the negative feedback aspect of discrepancy models (Archer, 1976a; Wiepkema, 1977), but positive reinforcement implies, in addition, positive feedback as a consequence of performance (or possibly a mechanism involving overcoming resistance). This would also fit with evidence that aggression shows warm-up or sensitisation effects (Chapter 10), i.e. its rate and level increase as a result of performance (Heiligenberg, 1974; Peeke, 1982).

Possibly, at lower levels of aggressive motivation, negative reinforcement predominates and behaviour tends to stop when the intruder has left. But positive feedback will still tend to maintain behaviour for some time after the intruder has left, the length of time depending on the level of motivation which has built up in the fight. Once this level exceeds a certain value, the balance will have changed so that positive reinforcement predominates owing to the strong positive feedback effects at higher motivational levels. Under these conditions, aggression will tend to outlast the presence of an opponent (it may become redirected), and animals may be observed looking in places where a fight previously occurred (Hinde, 1970; Tellegen & Horn, 1972). Nevertheless, in the absence of a target, aggressive motivation gradually diminishes since part of the signal supporting the positive feedback comes from the target, and in its absence the system will operate like a leaky bucket.

The hormonal consequences of winning or losing a fight can also affect behavioural responsiveness (Chapter 4). Defeat has been shown to decrease androgen levels in the swordtail (Hannes et al., 1984), the housemouse (Leshner, 1985), the rat (Schuurman, 1980), and the rhesus monkey (Rose et al., 1972; Bernstein et al., 1983). It increases corticosteroid levels in the swordtail (Hannes et al., 1984), the lizard *Anolis carolinensis* (Greenberg, 1983), the pig (Bouissou, 1983), and the housemouse (Leshner, 1980, 1983b). The second of these changes has behavioural consequences in mice (Leshner, 1980, 1981, 1983b) and probably in rats (Schuurman, 1980): it increases their subsequent submissiveness and defensive behaviour, thus augmenting the punishing effects of defeat (Kahn, 1951; Taylor, 1979).

Suppression of a rival's reproductive potential as a consequence of defeat was viewed as an important function of female aggression in mammals (Chapter 7). For example, Huck, Bracken & Lisk (1983) found that short periods of defeat by another female reduced pregnancy success in newly mated hamsters, and resulted in reduced litter size in pregnant ones. The physiological mechanism underlying this suppression is probably a stress-induced decline in reproductive hormones which parallels similar changes in defeated males (Leshner, 1975).

In lizards, body colour can change as a result of defeat, and Greenberg & Crews (1983) raised the possibility that this may be mediated by changes in the levels of epinephrine and melanocyte-stimulating hormone (MSH). Body colour is, in turn, a powerful stimulus for initiating attack in these animals (section 8.3).

Since a rise in testosterone levels may increase aggressiveness in a wide range of species, one possibility is that winning aggressive encounters may facilitate testosterone secretion, and hence increase subsequent aggressiveness. Wingfield (1985) has shown that plasma testosterone levels are elevated during the establishment and defence of a breeding territory in male song sparrows. Although no increase has been recorded in rodent studies (Schuurman, 1980; Leshner, 1983b), there does appear to be an increase in testosterone levels as a result of winning fights for several primate species (Rose *et al.*, 1972; Bernstein *et al.*, 1983). An interesting related human study is that of Mazur & Lamb (1980), who found that a competitive achievement (either winning a tennis doubles match for a cash prize, or receiving an MD degree) was associated with a rise in testosterone levels in men.

Testosterone levels are also increased in social situations which do not involve competition, including those associated with reproduction. There is evidence from birds such as pigeons, zebra finches and white-crowned sparrows that a significant part of the males' rise in androgens during the breeding season occurs as a result of social stimulation from receptive females (Harding, 1983; Runfeldt & Wingfield, 1985). In several mammalian species, including rats, mice, bulls, other ungulates and primates, the male responds to the presence of a receptive female with a rise in testosterone levels (Katongole, Naftolin & Short, 1971; Purvis & Haynes, 1974; Batty, 1978a, b; Bouissou, 1983; Bernstein *et al.*, 1983). This increase may facilitate aggression at a time when it is functionally appropriate, for example for mate-guarding (Parker, 1974a; section 7.5).

8.6 Conclusions and suggestions for future research

I began this chapter by describing the general design features of territorial aggression in vertebrates as a negative feedback loop in which the output (attack) would tend to remove the input (the intruder). On the input side, discrepancy-monitoring was regarded as a widespread principle, but there was also evidence for a matching process in some cases. The relative importance of the two requires further investigation. Other forms of competitive aggression included direct competition and frustration-induced aggression. Future research on the second of these should aim to extend the range of species and reinforcers investigated, to establish the generality of 'frustration' as a process underlying competitive aggression.

Internal control of the readiness to show aggression ensures that it occurs at appropriate times. For example, the influence of testosterone, which has a widespread facilitating effect in male vertebrates, enables increased aggressiveness to coincide with competition for reproductive resources. There appear to be several mechanisms underlying this influence. One of the most interesting for future research is the possible facilitation of attachment to a spatial area.

The following likely mechanisms underlying species and sex differences were considered:

1 changes in the motivational balance of aggression and fear to a given input;
2 the relative importance of matching and mismatching processes in analysing the input;
3 differences in the extent to which species internalise landmarks during territory formation;
4 hormonal influences.

Regarding the last of these, most research has concentrated on birds and mammals, with amphibians, reptiles and invertebrates being particularly neglected. Female hormones, whose influences may be more varied than those of the male, have also been little studied outside the mammals. Long-term influences of pituitary–adrenal hormones have, with one exception (Munro & Pitcher, 1985), been studied only in the mouse.

A socially restricted rearing environment prior to weaning leads to an increase in later aggressiveness in rodents such as the housemouse. This probably results from fewer opportunities to learn to gauge responses to the particular opponent. After weaning, restricted rearing may produce either heightened or diminished aggressiveness. It was suggested that the following developmental influences might be more profitable avenues for future research than further studies of restricted rearing environments:

1 the effect of intrauterine position in mice;
2 the father's influence in the preweaning period;
3 social influences on sexual maturation.

Concerning the first of these, vom Saal found that mice which had developed between two males were more aggressive as adults than mice from other intrauterine positions. He also suggested that this provides a mechanism underlying phenotypic polymorphism which is adaptive in fluctuating populations of microtine rodents; it would therefore be interesting to test this hypothesis in other microtine rodents which show population cycles, such as lemmings (Elton, 1942; Curry-Lindahl, 1962) and voles (Chitty, 1952; Godfrey, 1955).

The finding that the father's presence during the preweaning period facilitated adult aggressiveness in male mice raised the possibility of an olfactory imprinting-like mechanism, which might prove a fruitful topic for further study.

In considering mechanisms which produce behavioural adjustments in the short-term, we first considered a nectar-feeder which switches from territorial to non-territorial behaviour and back again according to nectar levels and pressure from intruders. Its behaviour was viewed in terms of choices between natural schedules of reinforcement, providing either a high return at a greater cost or a lower return at less cost. It would be interesting to try to simulate in the laboratory the precise schedules which may be controlling behaviour in the field.

Other short-term modifications of responsiveness included changes in aggressive behaviour as a result of previous aggressive interactions, mediated through either learning or hormonal mechanisms. We concluded that winning a fight can provide reinforcement in the absence of a conventional reinforcer such as food. This may involve either negative reinforcement (removal of an aversive stimulus), or positive reinforcement (involving appetitive motivation). It was suggested that the second process occurred only when aggressive motivation had been aroused, for example by a brief period of fighting, and that the nature of the reinforcement in any particular case depends on the balance between negative and positive feedback effects. This hypothesis also requires further investigation.

Further reading

Bekoff, M. (1981). Development of agonistic behaviour: ethological and ecological aspects. In *Multidisciplinary Approaches to Aggression Research*, ed. P. F. Brain & D. Benton, pp. 161–78. Amsterdam: Elsevier/North Holland.

Harding, C. F. (1983). Hormonal influences on avian aggressive behavior. In *Hormones and Aggressive Behavior*, ed. B. B. Svare, pp. 435–67. New York: Plenum.

Looney, T. A. & Cohen, P. S. (1982). Aggression induced by intermittent positive reinforcement. *Biobehavioral Reviews,* **6,** 15–37.

Taylor, G. T. (1979). Reinforcement and intraspecific behavior. *Behavioral and Neural Biology,* **27,** 1–24.

vom Saal, F. S. (1984). The intrauterine position phenomenon: effects on physiology, aggressive behavior and population dynamics in house mice. In *Biological Perspectives on Aggression,* ed. K. J. Flannelly, R. J. Blanchard & D. C. Blanchard, pp. 135–79. New York: Liss.

9

Fighting strategies: game theory models

9.1 Introduction

In Chapter 7, we were concerned with the functional question of whether it is advantageous to fight over a particular resource. In this chapter, we are principally concerned with whether it is advantageous to fight a particular opponent, and, if so, what strategy to use. The two issues do, however, overlap, since the strategy to be adopted (e.g. whether to 'escalate' a fight) depends partly upon the value of the resource. Nevertheless, the distinction is a convenient one, as it corresponds to two different functional approaches, the economic models discussed in Chapter 7, and game theory models, considered in this one.

Game theory models are an extension of the economic approach in that they are also concerned with calculating the costs and benefits of fighting. However, they introduce a number of further features which are necessary for a full understanding of the evolution of aggressive competition as a form of *social* behaviour. First and foremost, they are concerned with the way an animal fights, its fighting strategy. The strategy which will be optimal in a given set of conditions depends crucially on the behaviour of the opponent, or more precisely, on what other strategies are encountered, and how often.

Game theory is a branch of applied mathematics (von Neumann & Morgenstern, 1944), which enables the interaction of different strategies to be calculated by presenting them in a pay-off matrix as players in a game (e.g. Fig. 9.1).

The aim of most applications of game theory to animal conflicts is to determine the ESS or evolutionarily stable strategy (Chapter 1) for a given set of circumstances. A strategy is evolutionarily stable when no alternative one can replace it given the current conditions of the population (Maynard Smith, 1972). The ESS is calculated from the pay-off matrix by playing off the various strategies against one another. The exact result depends on:

1 the available strategies;
2 the costs and benefits they entail;
3 their frequencies.

In these calculations, it is generally assumed that every individual is equally likely to encounter every other one in the population. Obviously, this is a simplification, and departing from it will alter the results. Some recent models have, however, extended the game theory approach to non-random games (Toro & Silio, 1986).

In this account of the game theory models, they are described in non-mathematical terms, so that the precise reasoning behind many of the conclusions has been omitted, to concentrate on the conclusions themselves and whether or not they are supported by empirical evidence.

9.2 The hawk–dove game: the evolution of conventional competition

The simplest and most well-known game theory model of fighting strategies is the hawk–dove game (Maynard Smith, 1976, 1982) where there are only two strategies, 'hawks', which attack rapidly and inflict damage, and 'doves', which display, but retreat if the opponent attacks. Figure 9.1 shows the pay-off matrix when the cost of a serious injury (C) is a relative decrease in fitness of 100 units (-100), the time cost of a long dispute (T) is a decrease of 10 units (-10) and the relative increase in fitness for the winner (V) is 50 units ($+50$). The matrix shows that doves always lose to hawks (pay off $= 0$, since they retreat immediately), and hawks always beat doves (pay-off $= +50$, the benefit of victory). However, doves on average fare better in encounters with other doves ($+15$) than hawks do with other hawks (-25).

Figure 9.1 shows the pay-offs when there are equal numbers of hawks and doves. Suppose, instead, that the population consists of 10 per cent hawks and 90 per cent doves. There would then be only a low proportion ($\frac{1}{10}$th) of high cost (-25) encounters between two hawks, but a high proportion ($\frac{9}{10}$ths) of victories by hawks over doves ($+50$). Only 10 per cent of the doves' encounters would be against hawks, where they always lose (0); the other 90 per cent would be against other doves ($+15$). Calculating the fitness of the two strategies from their relative frequencies and the pay-offs shows that in such a population,

	H	D
H	$0.5\,(V-C)$	V
D	0	$0.5\,V - T$

	H	D
H	-25	$+50$
D	0	$+15$

Fig. 9.1. Pay-off matrix for simple hawk–dove game: hawk (H), dove (D), pay-off (V), cost of injury (C), time cost (T). Pay-offs all refer to gains in fitness from conflict, in arbitrary units. (It is assumed that animals with the same strategy will win 50 per cent of the encounters with one another.)

hawks would have a considerable advantage over doves. In a stable breeding population, where the two strategies are heritable, the proportion of hawks would gradually increase. As it did so, the proportion of their encounters which involved another hawk would also increase. Since these entail a high cost, the fitness advantage hawks have over doves at low frequencies would diminish. There would come a point when neither hawks nor doves would have an advantage over the other in terms of fitness. This point, which can be calculated graphically or algebraically, is evolutionarily stable. If a polymorphic population consisting of this particular ratio of hawks to doves is invaded by more animals of either type, the population would revert back to the stable ratio. Its precise numerical value depends on the costs and benefits of each type of interaction.

The same principle would apply if the population consisted of one type of individual which behaved like a hawk in some contests and like a dove in others (Maynard Smith, 1982). In this case, the ratio would refer to the proportion of occasions on which each strategy was adopted. This would be an ESS for that population, since any departure from it would be open to invasion by alternative strategies. Since, in this case, the ESS is a mixture of both strategies, it is termed a mixed ESS.

Animals which possess dangerous weapons refrain from damaging intraspecific fights, owing to the high risk of injury they would entail (Geist, 1974a: see Chapter 3). This is represented in the hawk–dove model by a high value for C, which, in the above example, biases the mixed ESS towards a pure dove strategy. Maynard Smith (1982) showed that in a simple hawk–dove model, where there is no time cost (T), the ESS is a mixed strategy defined by $P(H) = V/C$, where $P(H)$ is the proportion of time hawk is adopted. It follows that when the costs of injury (C) are relatively low, as occurs in animals without weapons, the hawk strategy will be more common, for example in toads (Davies & Halliday, 1978), newts (Smith & Ivins, 1986), and small (poorly armed) male thrips (Crespi, 1986). Alternatively, when the benefit obtained from victory is large (as in the case of protective or parental aggression: Chapters 3 and 5), the probability of adopting a hawk strategy is also high. This may apply even to animals where the costs of injury are great, such as red deer. Clutton-Brock *et al.* (1979) found that stags fought more frequently during the first 2 weeks of October, which coincided with the peak period of conceptions (Clutton-Brock *et al.*, 1982).

The basic hawk–dove model provides a starting-point for producing more complex models, for example by introducing additional strategies. Caryl (1981) added the 'prudent hawk' strategy, which attacks at the same level as a hawk but withdraws after a suitable period of time, even if no injury has occurred. He found that this strategy always wins over doves, and sometimes over hawks; its success depends on the riskiness of attack in relation to the value of the resource (i.e. C/V). Where the riskiness is great, for example where $C/V = 8$, prudent hawks form the majority (79 per cent) of the population. Caryl argued that this model seems intuitively to fit the behaviour of real animals better than the original hawk–dove model, where all escalated contests ended in serious injury. Caryl argued that contests are rarely carried on to the point where serious injury occurred. However, examples such as the red deer, where 6 per cent of rutting stags are severely injured each year (Clutton-Brock *et al.*, 1979), suggest we should be cautious in accepting this argument.

In the first well-known paper on game theory models of animal conflicts, Maynard Smith & Price (1973) introduced an elaborate model containing both hawk and dove (originally 'mouse'), together with bully, retaliator and prober-retaliator. A bully attacks initially but retreats if an opponent attacks; a retaliator attacks only if an opponent does so first; and a prober-retaliator attacks (low probability) or threatens (high probability) at first, then reverts to threat if the opponent retaliates, but attacks if it threatens (if it receives an attack, it attacks back). A large-scale simulation including all five strategies showed that retaliator was generally an ESS, but if there were more than 7 per cent doves in the population, it would be replaced by prober-retaliator. A simplified version of this model consisting of hawk, dove and retaliator (Maynard Smith, 1982) showed that retaliator was an ESS (with a polymorphic population consisting of equal frequencies of hawk and dove as an alternative one). All these simulations revealed the superiority of mixed, 'limited attack' strategies over the pure hawk or dove strategy. They illustrate how such conditional strategies could have evolved, particularly in animals with dangerous weapons.

The models described so far have been applied to the question of why animals with damaging weapons limit their use in intraspecific fights, and are particularly relevant to the question of the separation of competitive and protective forms of aggression in such cases. We should also expect a separation for parental aggression (Chapter 5) since losing would lead to a considerable decrease in fitness. For example, in the mantis shrimp, brooding females become considerably more successful in defending their brooding cavities than non-reproductive females (Montgomery & Caldwell, 1984). Their increased success arises mainly from a greater willingness to escalate a contest after being struck by a challenger, i.e. to become a 'conditional hawk' or retaliator.

The models introduced in this section are simple in their assumptions and consequently far-removed from reality. Nevertheless, they are useful for the following reasons (Krebs & Davies, 1981). First, they show that the pay-offs for any strategy are frequency-dependent, which shifts the emphasis from an optimal strategy to the concept of an ESS or stable strategy. Secondly, the exact ESS depends on which strategies are chosen for the game. Whether realistic strategies are chosen depends on the extent to which they match data from real animals. All of these simple models show that neither extreme strategy – hawk or dove – is stable except in extreme circumstances. The precise ESS always depends on the values assigned to the pay-off, and it will be difficult to provide precise numerical values for these. Nevertheless, it is possible to make broad qualitative predictions, and since the models involve *relative* fitness values, approximate numerical values can be assigned to them (Maynard Smith, 1982).

9.3 Ownership and the bourgeois strategy

The games described so far involve contestants which are all similar in fighting ability and in their position in the habitat. Most real animal conflicts involve contestants who are dissimilar to one another, either in fighting ability, or in the prior possession of a resource, or in the relative benefit obtained from the resource. In game theory models, these conflicts are referred to as asymmetric ones (Parker, 1974*b*). In this section, we consider one form of asymmetry, the possession of a resource or occupation of an area. This has been referred to as an 'uncorrelated' asymmetry since it is not logically connected to the fighting strategy of the contestants. In contrast, an asymmetry such as size (section 9.4) is referred to as a correlated asymmetry since it will affect the outcome and therefore is related to the fighting strategy.

Uncorrelated asymmetries have been modelled by Maynard Smith (1974, 1982) in a simple game involving hawks and doves together with the conditional strategy 'bourgeois', which responds like a hawk if it is the resident on an area (or owner of a resource), and like a dove if it is the intruder. In a population consisting of these three strategies, bourgeois will be the ESS. This shows that conventional acceptance of ownership comes to be used to settle contests when there is no correlated asymmetry or difference in the value of the resource to the two contestants.

At first sight, the model appears to have widespread applicability, for example to cases of territorial residency. However, in order to demonstrate that it applies in any particular case, the following predictions must be fulfilled:

  1 the 'owner' or resident must always win without a damaging fight (i.e. the intruder must not escalate);
  2 this result must be reversed for any given pair when ownership or residency is reversed;
  3 an escalated fight would occur if both animals perceive themselves to be the 'owner' of the resource or the resident (Maynard Smith, 1982).

Two studies which are commonly cited in relation to this model are those of Kummer, Gotz & Angst (1974) on intersexual bonding in hamadryas baboons, and Davies (1978*b*) on the transient mating territories of male speckled butterflies on the woodland floor. In both cases, contests are brief and their outcomes can be reversed depending on who is owner of the resource at the time (Krebs & Davies, 1981; Maynard Smith, 1982). In Davies' study, an escalated contest, which was damaging to both individuals, occurred when two butterflies were deceived into joint residency of a single patch of woodland. Here, the asymmetry on which disputes were usually settled was absent, and so both individuals behaved like hawks. Kummer *et al.* also observed escalated contests when two male baboons both perceived themselves as having access to the same female.

Sigg & Falett (1985) studied the possession of food in hamadryas baboons and found that a dominant partner would not take a food container or fruits from a subordinate one which already had possession; however, where pieces of food were thrown between the two animals, the dominant one would not allow the subordinate to gain possession. Two further experiments showed that the crucial feature was which animal uses a particular feeding place or a food can first, rather than a general preference for familiarity, suggesting that there is a concept of 'ownership' accepted by both animals which prevents them fighting over each resource item. The strength of this convention is seen in its ability to override established dominance relations. Although Sigg & Falett did not specifically relate their results to the game theory approach, their study provides the strongest evidence so far for the conventional bourgeois strategy described in Maynard Smith's model.

Returning to the question of whether the bourgeois model is relevant to territorial ownership, experimental studies on fish and rodents (Chapter 8) have shown that relatively short periods of familiarity with an area confer an advantage in subsequent encounters. Nevertheless, despite a superficial similarity to the bourgeois strategy, most cases of territorial ownership involve either a correlated asymmetry, or unequal pay-offs for residents and intruders, or both. In the first case, territories may have been obtained by larger animals (e.g. O'Neill, 1983, for *Philanthus* sp.; Harvey & Corbet, 1985, for dragonflies; Petrie, 1984, for moorhens; see also Table 9.1). In the second, the value of any but the most transitory territory will probably be greater to its owner than to a challenger, since the owner will have already learned about features such as food sources and refuges (see Davies, 1981, for an example in a field study of pied wagtail territories). Thus, even in the relatively featureless habitats used in experimental studies of territory formation, the resident could possess such advantages as a faster reaction to danger, after only a short period of time. In laboratory studies of fish and rodents (e.g. De Boer & Heuts, 1973; Thurmond & Lasley, 1979), territorial ownership developed over a period of at least 16 hours, presumably by a process of exposure learning (Sluckin, 1972). In such cases, residents would possess advantages derived from their knowledge of the area. This applies to all cases where residence is established only after a period of time, including some studies where the results were specifically related to the bourgeois model (e.g. Yasukawa, 1979). There are, however, cases where a new resident behaves like a territory-holder almost immediately. This occurred in Davies' study of the speckled butterfly, where ownership was apparently established according to the order of landing on an area, rather than by spatial learning. Hansen & Rohwer (1986), studying red-winged blackbirds, also observed that 'floating' males which took over a vacant territory behaved like a territory owner only minutes afterwards.

9.4 Resource holding power

As we noted in the previous section, in most animal conflicts, one contestant will have an intrinsic advantage such as greater size or better weapons or more experience of fighting. Parker (1974*b*) referred to this as a correlated asymmetry since it will be related to the outcome of the fight. He argued that the contestants would assess their opponent's fighting ability – or 'resource holding power' (RHP) – relative to their own, and their behaviour would vary according to this assessment.

Parker (1974*b*) analysed correlated asymmetric contests using game theory, and showed that strategies involving escalation would be restricted to opponents whose RHP is closely matched (but see the war of attrition model in the next section). Where it is not, the animal with the largest RHP will win since the opponent withdraws after a conventional display.

Size is probably the most important indicator of RHP which leads to an asymmetric contest. Table 9.1 shows its influence on the initiation, duration and outcome of encounters for a wide variety of animals: generally, larger animals possess an advantage, which is used to settle disputes. Where size is equal, contests do tend to last longer (Austad, 1983; Suter & Keiley, 1984, for bowl and doily spiders; Clutton-Brock *et al.*, 1982, for red deer), and involve more intense interactions (Robinson, 1986, for yellow-rumped caciques; Turner & Huntingford, 1986, for a cichlid).

Weapons may be more important than size as an RHP indicator in other cases, or else used as an additional factor where the size difference is small (e.g. Geist, 1966, for Stone's sheep; Neil, 1985, for hermit crabs).

A third form of RHP assessment is based on information about earlier encounters with that opponent (Van Rhijn & Vodegel, 1980). This is obviously important in social groups with individual recognition, and will reflect previously established dominance relations (Chapters 7 and 8). Thouless & Guinness (1986) report that female red deer use previous experience in this way to resolve conflicts between familiar individuals.

Parker's model predicts that animals will assess their opponent's RHP before entering into a fight. One way in which they could do this would be to begin to fight and then withdraw when it becomes apparent that

Table 9.1. *The influence of size on the initiation, duration and outcome of aggressive encounters*

Taxonomic group	Species	Behaviour	Reference
Class: Hydrozoa	*Actinia equina* (sea anemone)	Larger animal initiates aggression sooner, and is more likely to win	Brace & Pavey (1978)
Class: Amphineura	*Lottia gigantia* (giant limpet)	Resident dislodges smaller intruder	Stimson (1970)
Class: Gastropoda	*Hermissenda crassicornis*	Larger animals initiated encounters and tended to win	Zack (1975)
Class: Crustacea	*Erichthonius braziliensis* (amphipod)	Only larger intruders evict smaller tube-dwellers	Connell (1963)
	Clibanarius vittatus (hermit crab)	Large animals more aggressive and less fearful, and dominant over smaller ones	Mitchell (1976)
	Pagurus bernhardus (hermit crab)	Larger crab initiates encounter in large majority of cases. Relative size good predictor of outcome when size difference was sufficiently large	Dowds & Elwood (1983) Neil (1985)
	Carcinus maenas (green shore crab)	Larger males win fights over females	Berrill & Arsenault (1982)
Class: Insecta	*Diapheromera veliei* and *D. covilleae* (stick insect)	Larger males win fights over females	Sivinski (1978)
	Gryllus integer (field cricket)	Heavier males more aggressive than lighter ones	Dixon & Cade (1986)

(continued)

Table 9.1 *continued*

Taxonomic group	Species	Behaviour	Reference
Class: Insecta *(continued)*	*Philanthus crabronifermis* and *P. pulcher* (beewolves)	Larger males win encounters over territory. If removed, smaller males replace them	O'Neill (1983)
	Pyrrhosoma nymphula (dragonfly)	Larger male adults win more territorial disputes and obtain more matings near water	Harvey & Corbet (1985)
	Plectrocnemia conspersa (caddis-fly larvae)	Body size determines outcome of disputes over nets	Hildrew & Townsend (1980)
Class: Arachnida	*Nephila clavipes* (orb weaving spider)	Larger males assume hub position near female (feeding advantage)	Christenson & Goist (1979)
	Agelenopsis aperta (funnel-web spider)	Larger male wins significantly more often	Reichert (1978)
	Frontinella pyramitela (bowl and doily spiders)	Larger male wins 75% of all male–male interactions.	Suter & Keiley (1984)
		Encounters between well-matched males longer	Austad (1983)
Class: Osteichthyes	*Lepomis macrochirus* (bluegill sunfish)	Large residents show more aggressive responses to both large and small intruders than small ones	Henderson & Chiszar (1977)
	Betta splendens (Siamese fighting fish)	Larger body size confers an advantage in agonistic encounters	Bronstein (1984)

Taxonomic group	Species	Behaviour	Reference
	Oreochromis mossambicus (Mozambique mouthbreeder)	Larger fish wins fight. Contest intensity decreases as size increases	Turner & Huntingford (1986)
Class: Amphibia	*Uperoleia rugosa* (Australian frog)	Heavier males usually win fights	Robertson (1986)
Class: Reptilia	*Vipera berus* (viper)	Larger males win encounters. Encounters between well-matched animals last longer	Kelleway & Brain (1982)
	Anolis sagrei (brown anoles)	Larger males defended perch sites better than smaller ones; also showed more challenge displays and most often entered other males' territory	Tokarz (1985)
Class: Aves	*Gallinula chloropus* (moorhen)	Heavier birds won contests in winter flocks and established territories	Petrie (1984)
	Cacicus cela (Peruvian yellow-rumped cacique)	Heavier birds dominant over smaller ones. Encounters between similar-sized birds more intense	Robinson (1986)
Class: Mammalia	*Mesocricetus auratus* (golden hamster)	Males dominate females only when much heavier	Marques & Valenstein (1977)
	Rattus norwegicus (wild Norway rats)	In encounters between unequal-sized rats, the larger attacks and the smaller escapes	Robitaille & Bovet (1976)

the opponent has a higher RHP. However, this would entail high risks and incur heavy fitness costs. Parker's model indicates, therefore, that animals should make assessments *before* beginning a fight, so as to avoid these possible costs. He predicted that animals will perform behaviour patterns providing reliable indicators of relative RHP early in an aggressive encounter, that these will be used by the opponent for assessment, and that they will affect its propensity to attack so that contests are settled without escalation.

Maynard Smith (1982) applied a modified hawk–dove game to such cases where the contestants initially possess knowledge of each other's RHP. He introduced a third strategy 'assessor' (A) which chooses hawk if its RHP is greater, and dove if it is smaller. In conditions where escalation is costly, A is shown to be an ESS: this applies when the cost of the assessment phase is less than the cost of losing an escalated fight $(c < C)$, and even if the perceived RHP is not a perfect predictor of which animal would win. It was also shown that only a pure ESS (rather than a polymorphic population) can exist in such cases. However, in conditions where assessment is costly and escalation less dangerous, the hawk strategy was shown to be the ESS.

The assessor strategy can be illustrated by Riechert's (1978, 1984) study of encounters between female funnel-web spiders at their web sites. The spiders initially assessed each other's weights, and their subsequent activities depended on the results of this assessment. If a spider had a large weight advantage, it would escalate directly to higher-intensity (and higher-cost) attacks. If it was smaller (and was an intruder), it would retreat. In this case, residence is a complicating factor, since a smaller resident will adopt a retaliator strategy (section 9.2), and would then be likely to lose the escalated contest.

At this point, we need to refer to a distinction which has been made between signalling information about RHP, and signalling about motivation or intentions. The concept of RHP refers to features such as size, strength or weapons which are difficult to imitate and costly to acquire in the first place (Krebs & Davies, 1981; Maynard Smith, 1982). On the other hand, false information about motivational state or intentions can easily be transmitted. To do so would cost little, since it would involve a transient signal rather than an increase in size or the production of a weapon. Maynard Smith (1982) argued, therefore, that there is no functional reason why signals about intentions should be accurate. (We discuss this issue further in section 9.5 and in Chapter 10). Signals for RHP, however, are more likely to be accurate. Nevertheless, there may be some conditions under which RHP assessment is inaccurate. Maynard Smith & Parker (1976) set out to model such conditions using the hawk–dove–assessor game. Essentially, this analysis concerned the effect of mistaken assessments on the outcome. Where an assessment cue provides an uncertain predictor of the outcome of an escalated contest, the cue can still be used to settle contests without escalation, unless the injury cost of escalation is minimal. Where there is uncertainty about the cue itself, escalated fights will occur more frequently; in a population of assessors they will occur when the contestant with the smaller RHP mistakenly estimates that it has the larger RHP (Maynard Smith, 1982). Despite this, the assessor strategy is still stable under a wide range of conditions, and escalated contests will generally occur only infrequently.

Although we previously suggested that it is difficult for an animal to fake RHP signals, there are nevertheless many cases of bluff, where an animal increases its apparent body size in aggressive encounters without altering its RHP. Maynard Smith & Parker (1976) also investigated this question in a game theory model and concluded that bluff would be effective and would spread in a population only if it could not be distinguished from a real signal except during an escalated contest. However, if individuals are able to develop the capacity to distinguish bluff from reality under most conditions, their strategy will be to ignore the bluff and follow the actual RHP. Maynard Smith & Parker concluded that in some circumstances, such as when there is a possibility of serious injury, an ESS involving extensive bluff can evolve.

Although features such as size and weapons can be developed only over a period of time and at considerable cost, there is one way of circumventing this which occurs in social primates such as male baboons (Packer, 1977), which is to enlist another male's help in a temporary coalition to fight a rival. Since this seems to be based on the principle of reciprocal altruism (Trivers, 1971), it would entail the additional cost of aiding the altruist when help is required. In such coalitions (which have also been reported among chimpanzees: de Waal, 1984), the RHP of the instigator is considerably increased without the long-term costs of developing additional body mass or weapons. Of course, such ways of developing additional RHP, at a reduced cost, have been vastly extended in human conflicts, with organised fighting groups and weapons. Once RHP is emancipated from the costs involved when it is tied to a permanent part of the animal's body, and it becomes subject to cultural evolution, its evolutionary development can progress at an escalated rate.

9.5 The war of attrition model and the evolution of aggressive displays

In the previous section, we distinguished between transmitting information about RHP, which is difficult to fake, and transmitting information about fighting intentions, which is much easier. Game theory has also been used to analyse this second case. Maynard Smith (1974) introduced a model referred to as 'the war of attrition'. In this, he considered a contest between two equally matched doves, i.e. animals which would not escalate, the winner being the one which is prepared to display for the longest time. Maynard Smith's model shows that no pure strategy is evolutionarily stable, since any population of animals playing strategies of a fixed value will be vulnerable to invasion by strategies paying slightly higher. He calculated that the ESS is a mixed strategy, given by the following distribution:

$$p(x) = I/v \exp{(-x/v)}, \tag{9.1}$$

where v is the pay-off to the victor, and x is the length of the display.

He also showed that an evolutionarily stable population will either be genetically polymorphic or consist of individuals whose behaviour varies from contest to contest according to this negative exponential distribution. The important conclusion for the evolution of aggressive displays was that selection would oppose any tendency for the intensity of display to reveal future intentions. Suppose the population consisted of animals which signalled their intended display length honestly at the beginning of the contest. It would pay such individuals to retreat when they perceived a signal indicating a longer display time than their own. But such an honest population would be vulnerable to invasion by individuals whose signals exaggerated their intended display times and therefore would not be evolutionarily stable (Krebs & Davies, 1981).

As Krebs & Davies (1981) pointed out, the original war of attrition model is unlikely to hold in many circumstances since most conflicts will be settled by asymmetries. In practice, even if it is found that display lengths* are distributed according to the negative exponential shown in equation 9.1, this does not necessarily mean that the war of attrition model applies (Maynard Smith, 1982; Turner & Huntingford, 1986). A study by Parker & Thompson (1980) on struggles between male dung-flies – arising from mate-guarding – illustrates this point. Even though a superficial concordance was found between the observed persistence

---

* Display *lengths* are distributed exponentially but with the mean V/2, the average pay-off, instead of v (Maynard Smith, 1982).

times and those predicted by the war of attrition equation, further analysis revealed that the contests were asymmetric ones, suggesting that the struggles were being settled by an RHP assessment process.

Sigurjonsdottir & Parker (1981; subsequently confirmed this impression. Strategies were related to the size of the intruder relative to the one guarding the mate (resident). As the disparity between them decreased, so did the persistence of the intruder, and an attacking intruder was invariably larger than the resident. Other factors, connected with resource value, such as the size of the female, and the time the male had spent guarding, also contributed to the persistence of the intruder and the resident, respectively. Presumably, the interaction of these various factors resulted in a random distribution of persistence times.

There are also limitations in the prediction from the original war of attrition model that contest lengths will follow the negative exponential distribution. Norman, Taylor & Robertson (1977) showed that this distribution would be obtained only if the cost of displaying was a linear function of duration. As Caryl (1979, 1981) pointed out, this is not necessarily the case, for example when different displays vary in intensity. In order to produce a more general model, Caryl (1979) introduced a cost function, $q(x)$, relating the cost of displaying to its duration, x. Since the cost of displaying will involve features such as the cost of neglecting other behaviour, and energy expenditure, it will not necessarily be a linear function of duration even when the intensity of displaying is constant. Norman *et al.* showed that in these circumstances, the ESS will be given by the following distribution:

$$p(x) = I/v \; q^1(x) \exp\left[-q(x)/v\right], \tag{9.2}$$

where $q(x)$ and $q^1(x)$ are the cost function and its differential, respectively.

The difficulty with this more general model is that there are few data available on cost functions with which to test it (Caryl, 1979).

In the original war of attrition model the population consisted of 'doves'. In other words, it was assumed that the contests would be settled without injury. As Caryl (1981) pointed out, real animal conflicts often involve 'escalated' displays. An escalated display – in the sense used by ethologists – is a progressive series which becomes more costly, in terms of both energy expended and risks taken. (Caryl pointed out that the term 'escalation' has often been used in game theory to denote a transition from one discrete category, such as a dove, to another, such as a hawk.) Bishop & Cannings (1978) proposed an extension of the original war of attrition model which would incorporate the possibility of injury to one of the opponents. In this model (referred to as 'the war of nerves' by Caryl, 1981), it is assumed that the injured animal will stop at once. The ESS is to play the following negative exponential.

$$p(x) = (I/M) \exp (-x/M), \tag{9.3}$$

where x is the length of the display and M is the mean, given by the following:

$$M = v/[\mu - \lambda(v - D)], \tag{9.4}$$

where μ is the cost per unit time of the display, λ is the chance of injury, v is the pay-off and D is the cost.

The model predicts that the greater the severity of a possible injury, the more likely the contest will be to end without injury to either opponent. In these circumstances, most contests will involve a low cost, but a few will entail a high cost. The variance in pay-off over a series of fights would therefore be greater when a more severe injury is possible (see Caryl, 1981, for further discussion of this point).

The major impact of the war of attrition models has been in relation to their prediction about the nature of the communication during aggressive displays, and we discuss this in the next chapter. However, the basic war of attrition game has also been used in more complex models involving the interaction between ownership, RHP and resource value. These are considered in the next section.

9.6 Interaction between ownership, RHP and resource value

We have already considered the influence of ownership and of RHP on fighting strategies. A third form of asymmetry is resource value, which may be different for two contestants, especially between the owner of a resource and an intruder (section 9.3). In this section, we consider models involving the interaction of these different asymmetries.

It was pointed out in section 9.3 that the value of a territory is often greater to its owner, owing to advantages derived from learning about its features (e.g. Davies, 1981). Maynard Smith (1982) applied the hawk(H)–dove(D)–bourgeois(B) game to cases where the value of the resource was greater to the owner (V) than to the intruder (v), and he concluded that B was usually the ESS. In this game, it is assumed that the RHP of the contestants is equal and they have complete information about the value of the resource to one another.

Hammerstein (1981) also used the hawk–dove game to consider the case in which there is an asymmetry in both ownership and RHP, again assuming that each contestant had perfect information about the other. He concluded that when the risk of injury through escalation (*c*) was considerable compared to the value of the resource (*V*), contests may be settled by uncorrelated asymmetries (such as ownership) even if a correlated asymmetry (such as size) exists. However, if the correlated asymmetry exceeds a critical value, it will be used to settle the dispute. Consequently, an intruder with sufficient advantage in fighting ability will ignore ownership, resulting in cases where a smaller owner engages in nearly hopeless fights against escalating intruders.

Hammerstein and Maynard Smith (1982) have both cited the study of Hyatt & Salmon (1978), on fights over burrows between male fiddler crabs, in relation to this model. In about 90 per cent of cases, the owner won the contest, but in the majority of other cases, the intruder was larger. It seems that the disputes were generally settled by ownership, but that a sufficiently large size difference could override this. Very similar results have since been obtained for the red spotted newt by Smith & Ivins (1986). However, neither study rules out the possibility that a difference in resource value associated with ownership, rather than ownership *per se*, settled the dispute.

A second study which has been presented in support of Hammerstein's model is Riechert's analysis of encounters between adult female funnel-web spiders at web sites (Riechert, 1978, 1979). In this case, residency did not have an overall effect on the outcome. Size was the most important factor, the larger spider winning 91 per cent of encounters. Where body weights were similar, however, the resident was more likely to win. Whenever a resident lost, the fight was an escalated one, with a high energetic cost. Although this is apparently consistent with one of Hammerstein's predictions (see above), there is the added complication that residents but not intruders possess knowledge of the resource value, and this may account for their persistence. In a study of contests between male bowl and doily spiders over access to receptive females, Suter & Keiley (1984) also found that larger spiders won the majority of encounters. Again, when their sizes were closely matched, a clear advantage of residency was apparent. However, Austad (1983) found that this was an apparent advantage of residency only because the value of the resource (not measured in Suter & Keiley's study) was greater to the resident.

Two conclusions are apparent from these studies. The first is that in two species of spiders (but not in fiddler crabs or the red spotted newt), size overrides ownership in determining the outcome of territorial encounters. A similar finding has been reported more widely (Table 9.1): for example, in caddis-fly larvae, wasps and moorhens, heavier animals won territorial encounters irrespective of prior residence (Hildrew & Townsend, 1980; O'Neill, 1983; Petrie, 1984). It is clear that these findings do not support Hammerstein's conclusion that contests can be settled by an uncorrelated asymmetry even if a correlated asymmetry exists. However, without precise measures of the other factors in his model, i.e. relative RHP, injury and resource value, it is difficult to tell whether these examples are indeed inconsistent with his predictions.

The second conclusion is that in all cases where contests were decided on the basis of ownership, rather than RHP, unequal pay-offs rather than the bourgeois strategy were probably responsible. We must conclude, therefore, that Hammerstein's model is unlikely to apply widely to territorial residency because ownership will usually involve unequal resource value (section 9.3). In an earlier section (9.3), we described Sigg & Falett's (1985) study of food possession among baboons. They showed that a dominant male hamadryas baboon would not attack a subordinate partner which possessed a food container or a piece of fruit. However, when ownership was not established, the dominant took the food on each occasion. These circumstances do fit the conditions for Hammerstein's model, namely that the risk of injury through escalation should be considerable compared to the value of the resource. The observation that among females (where the risk of injury is smaller), the dominant member of a pair is more likely to ignore ownership and to take the food is also consistent with Hammerstein's model (Archer, 1986c).

Hammerstein's approach to modelling interactions between asymmetries assumes that each contestant has perfect information about the other: this enabled the dichotomous categories of 'escalation' and 'non-escalation' to be used. These assumptions are unrealistic not only in view of the gradual nature of escalation in animals fights (Caryl, 1981), but also because animals often do not have complete information about their opponent, especially how valuable the resource is to it, and on some occasions its RHP as well. In the previous section, we considered a study by Sigurjonsdottir & Parker (1981) on male dung-fly struggles over receptive females. In this, it was found that a recently mated male (the 'resident') possesses information about the value of the female which is not available to an attacker. Similar asymmetries in information availability have also been found in male spiders fighting over a mate (Riechert, 1979, 1984; Austad, 1983).

In section 9.4, we referred to the 'assessor' strategy (Maynard Smith & Parker, 1976), which entails a phase in which the opponent's RHP is assessed, implying that such information may not be readily available at the beginning of the contest. The general approach to modelling cases where information about asymmetries is uncertain has been to use the war of attrition game. Bishop, Cannings & Maynard Smith (1978) extended the basic model to cases where the owner of the resource knew its value, but the intruder knew only the distribution of the resource population from which it came (as in the examples cited above). The ESS in this case was for the owner to choose persistence times whose means are directly proportional to the value of the resource. Thus, an owner defending a resource of little value will be less persistent than one defending a valuable resource.

Parker & Rubenstein (1981) developed the war of attrition model further, to consider interactions between RHP and resource value (see also Hammerstein & Parker, 1982). They pointed out that such interactions could be either non-contradictory, where the opponent with the highest RHP (or residency) also has the most to gain, or contradictory, where the one with the highest RHP has the least to gain. The first case is straightforward and will follow the rules discussed in the section on RHP (section 9.4). The assessor strategy will be an ESS when the cost of assessment is small, escalation is dangerous and the RHP signal a good predictor of victory.

Contradictory interactions have been considered in terms of whether an animal is occupying a particular 'assessment role'. Earlier, Parker (1974b) had proposed that the individual with the lesser score for V/K (where V is the resource value and K is the rate at which costs are incurred) should retreat under asymmetrical conditions. He showed that a convention based on this value is the only ESS where opponents can assess their roles accurately before beginning a contest, and where the costs of an escalated contest would rise continuously at a fixed rate.

If we consider two individuals with resource values Va and Vb, during a contest the costs will rise continuously but asymmetrically for the two individuals (represented by Ka and Kb). If $Va/Ka > Vb/Kb$, an assessor strategy based on the rule 'withdraw if in role b, persist in role a' will be an ESS in a population that ignores the asymmetry. Furthermore, Parker & Rubenstein found that the assessor strategy has to show only a greater than chance success in assessment for it to spread in the population. Thus, even if the animals are poorly informed about whether they occupy role a or b, the tendency to drop out of contests first, when in role b, will spread.

One study which may be relevant to this is that of Krebs (1982), who removed great tits from their territories and released them after replacement pairs had occupied the spaces. He found that fights between replaced residents and their replacements were escalated compared with those between residents and intruders or established neighbours. At first sight, this seems to correspond to the bourgeois game described in section 9.3. However, Krebs also found that the likelihood that a contest would be escalated and that the former owner would win increased over a period of several days and then diminished.

Krebs argued that since reversal of ownership and the degree of escalation occurred over a period of days, the advantage of ownership must be due to a pay-off asymmetry rather than to an arbitrary convention (or an RHP asymmetry). He suggested that the assessor rule applied and that the bird with the greater value for V/K retreats.

In most cases, establishing a territory follows the gradual pattern found in Krebs' study (section 9.3): residents which are removed and replaced within minutes or hours tend to show an advantage over their replacements. Parker & Rubenstein's assessor rule may, therefore, have wide applicability. One problem in testing this is that it is difficult to specify the value of a resource to the territory-holder and intruder. The general assumption is that after a period of spatial learning, resources will be of greater value to the resident, since an animal on a familiar area may be more successful in a variety of activities, including avoiding predators, copulation or feeding. But it is difficult to measure this. Davies (1981) did, however, demonstrate that the winter feeding territories of pied wagtails were more profitable in terms of feeding rate for owners than for intruders.

Parker & Rubinstein's model can be tested more satisfactorily in those cases where a resource has a specific value in reproduction. In many cases of mate-guarding, only the resident male will possess information about the reproductive state of the female (resource value), since an intruder will not know how long the resident has been guarding. This is the case for the dung-fly struggles, described in section 9.5.

The most complete experimental test of Parker & Rubinstein's model is Austad's (1983) study of contests between male bowl and doily spiders over access to a female. Male spiders vigorously defend a female with whom they have not yet finished mating, and they engage in grappling fights on the webs. Since mature spiders rarely eat, but wander around in search of mates, these fights represent fairly direct competition for reproductive success. There is little sperm displacement as a result of a male takeover, so a resident male stands to lose only those fertilisations not yet achieved when the fight begins. There is no evidence in this species that males can assess the relative number of eggs contained in females of differing body size. These two features enable the value of the female for the resident and the intruder to be estimated from the fertilisation rate and the expected number of fertilisations obtainable by foraging elsewhere.

Austad manipulated the value of the female (Vr) by introducing the intruder to the resident at five different points:

1 simultaneously;

2 1 minute later, when the resident had copulated for 1 minute but had not transferred any sperm (at this point the resident cannot tell whether the female had previously mated);

3 at the end of the pre-insemination period, when the male will have determined that the female is a virgin adult and will be ready to discharge sperm;

4 7 minutes after insemination has begun, at which point 92 per cent of the total fertilisable clutch will have been fertilised;

5 after 21 minutes when 99 per cent will have been fertilised.

Austad also manipulated RHP in these fights, by arranging them into encounters between the following:

1 individuals which were of similar length (an accurate predictor of outcome);

2 those which differed slightly;

3 those which differed greatly.

He then calculated an estimate of the costs of fighting, in terms of the expected reduction (through the possibility of injury) in lifetime reproductive success per minute of combat, for the three different size categories.

The results provided evidence of an assessment strategy in that grapple duration was inversely correlated with size difference for males introduced simultaneously: lack of persistence by the loser determined this result. When the combatants were similar in length, the resident won more often when its expected gain was higher than that of the intruder, and least often when it had a lower expected gain.

It was argued that intruders would not differ according to when they were introduced (since they cannot tell the female's value). Thus, any differences which are a function of the time an intruder is introduced should reflect changes in the persistence of the residents resulting from changes in resource value. Indeed, the longest contests occurred when the resident had most to gain, and the shortest when there was least to gain. Austad noted how persistent smaller residents were when contending for a female of high value. Fights in this category were so severe that 90 per cent of them ended in a fatal or disabling injury for one of the participants. This result is consistent with the prediction from the model of Bishop *et al.* (see above) that an owner will choose persistence times proportional to the value of the resource; it is also supported in other studies on hermit crabs (Dowds & Elwood, 1983) and funnel-web spiders (Riechert, 1979, 1984).

Austad calculated values for V/K, for the various treatment categories, to determine whether this could be used to predict contest length in the manner indicated by Parker & Rubinstein. The nearer the values of V/K for the two contestants, the longer the predicted grapple durations. For four treatments this was the case, but for another two it was not. However, it was found that in these two conditions, and in these only, each contestant perceived the other's value of V/K to be larger, i.e. a result which would shorten contest duration. There was also one condition involving unequal-sized opponents where each combatant would perceive its own V/K value as larger: as predicted, contest durations were significantly longer than for others involving unequal-sized opponents, and the majority of fights involved serious injury.

Although Austad's study is the most precise test of Parker & Rubenstein's model, and provides convincing support for the assessment rule, there are other studies which provide further, albeit less detailed support. Ewald (1985) experimentally investigated the interaction of resource value and RHP in hummingbirds. RHP differences were represented by using adult and juvenile birds, the former usually dominating the latter. However, this was reversed when there was an asymmetry in expected pay-off. By training pairs of birds on adjacent territories to feed from feeders dispensing different rates of food, and then gradually merging the two territories, Ewald found that if a juvenile which expected a high rate of food presentation on its territory was paired with an adult which expected a low rate, the juvenile would be likely to win the contest, thus reversing the outcome when both birds expected the same pay-off. Although not explicitly designed to test Parker & Rubinstein's model, the results of this study again support, in general terms, the V/K rule.

In the studies reviewed so far, it is assumed that the different asymmetries act at the same time. In some conflicts, such as those of the hermit crab, *Pagurus bernhardus*, the contestants gradually gain information about asymmetries as the encounter progresses. Dowds & Elwood (1983, 1985) instigated encounters between crabs of different sizes (small or large), occupying preferred or non-preferred shells, and found evidence for a series of assessments as the encounter progressed. The initial decision whether or not to engage in conflict depends on RHP assessment from a distance, the larger crab usually initiating the encounter. At this stage, the quality of the initiator's shell plays only a small part, and the quality of the opponent's shell none at all. Whether to escalate the conflict depends on several factors:

1 a more accurate assessment of RHP, which generally leads to withdrawal if a smaller crab had initiated the conflict;

2 an assessment of the quality of the defender's shell, an initiator being more likely to escalate if this is from a preferred species;

3 the quality of the initiator's shell, escalation being more likely if this is from a non-preferred species.

Dowds & Elwood found that the initiator used information from the last two assessments to compare the quality of its own shell with that of its opponent, i.e. to estimate the expected pay-off, and to escalate if this was high. A defender, on the other hand, only has information about its own shell, and will defend a good-quality shell more vigorously than it will a poor-quality shell.

9.7 Conclusions and suggestions for future research

Models such as the hawk–dove game addressed the issue of how conventional ('limited war') strategies could have evolved through individual selection. The ESS was shown to be a mixed hawk–dove strategy whose composition depends on the costs of fighting (C) and the value of the resource (V): for example, when $V > C$, the hawk strategy becomes more pronounced. When a conditional strategy, 'bourgeois', which responds as a hawk if it is the owner and as a dove if it is the intruder, is added to the hawk–dove game, bourgeois becomes the ESS. It was argued that this model had only a limited application to territorial ownership, where the apparent advantage of residency generally results from an asymmetry in resource value or size between the opponents.

Parker's RHP model concerns contests between opponents which differ in features such as size or weapons. An 'assessor' strategy, which withdraws if its own RHP is the smaller, is an ESS whenever escalation is costly. Since RHP signals are tied to bodily structures, they are difficult to fake: bluff can only evolve when the imitation looks like the real signal and fighting would entail the possibility of serious injury if the signal were real.

The war of attrition model involved equally matched doves. The ESS is a mixed variable strategy whose duration follows a negative exponential distribution and which gives no reliable information about future intentions. Unfortunately, the same distribution can be produced in some asymmetric contests, thus limiting the model's predictive power. Refinements of the model have either replaced the duration of display by a cost function or incorporated the possibility of injury.

The interaction of RHP, ownership and resource value was considered in more complex models. One, derived from the hawk–dove game, predicted that where the cost of escalation (C) is high compared to resource value (V), ownership may be used to settle a dispute even though an RHP difference exists. This seemed to occur when animals with dangerous weapons contested food items; however, in most territorial disputes, resource value differences or size appeared to be the crucial variables.

Where there are inequalities in the information the contestants possess about the resource value, a strategy leading to persistence in fighting for a contestant with an apparently higher value of V/K (V = resource value; K = a cost function) was shown to be an ESS in a model based on the war of attrition game. A carefully controlled experimental study of contests between male spiders supported many of the predictions of this model (see below).

In this chapter, we considered both game theory models and studies relevant to their predictions. The models themselves provide a rich source of hypotheses for empirical testing, and will surely stimulate further research. Many of the existing studies used in game theory discussions were not designed to test their predictions. This highlights the need for future empirical studies to be designed to test specific game theory predictions, rather than adopting the easier approach of restricting them to providing *post hoc* explanations. There is one existing study which provides a model for future predictive tests. In a study of conflicts between male spiders, Austad (1984) carefully controlled variables such as resource value, ownership and RHP in an experimental design which was able to reveal powerful support for the predictions of one game theory model (section 9.6).

The models themselves will undoubtedly be further refined in the future. Essentially, all those reviewed in this chapter were derived from a modification of the hawk–dove and war of attrition models. One approach apparent in the development of game theory models over the last 15 years has been for models to involve more 'realistic' features, including additional conditional strategies, non-random games (Toro & Silio, 1986), and other more complex assumptions. In addition to a continuation of this trend, we should also expect new forms of model in the future, covering such topics as territory formation (Maynard Smith, 1982) and repeated encounters (Caryl, 1981).

Game theory models of conflicts are an unusual area of animal behaviour research, in that theory and empirical research relevant to testing the theoretical predictions have developed separately. In future, it would be advantageous for greater integration between theory and empirical work to be attempted.

Further reading

Austad, S. N. (1983). A game theoretical interpretation of male combat in the bowl and doily spider (*Frontilla pyramitela*). *Animal Behaviour, 31*, 59–73.

Caryl, P. (1981). Escalated fighting and the war of nerves: game theory and animal combat. In *Perspectives in Ethology 4*, ed. P. P. G. Bateson & P. Klopfer, pp. 199–224. New York: Plenum.

Maynard Smith, J. (1982). *Evolution and the Theory of Games*. Cambridge & New York: Cambridge University Press.

10

Mechanisms of fighting strategies

10.1 Introduction

The game theory models reviewed in Chapter 9 are concerned with the costs and benefits of entering into a conflict. Essentially, they are evolutionary models, but it is also supposed that both costs (such as those resulting from the opponent's fighting ability) and benefits (the resource value) are assessed prior to or during the course of the contest, and that the result influences the degree of escalation and who wins the encounter. Game theory models therefore make predictions related to motivational questions in addition to being concerned with function. Consequently, some of the models can be used as starting points for analysing motivational decision making in contests.

In this chapter, I first describe in general terms the ways in which the benefits and costs involved in game theory models are manifested in the immediate control of behaviour. Benefits involve various forms of resource, which were considered in Chapter 8, and the costs are related to RHP assessment. This is followed by consideration of a specific motivational model (by Maynard Smith & Riechert, 1984) which simulates the decisions made during a fight by two spiders. The model uses functional concepts such as RHP and resource value, and it illustrates a rather different approach to modelling aggressive behaviour from that described earlier in the book (Chapter 4).

Escalation – the sequential replacement of lower- by higher-level acts as the fight progresses – is a crucial feature of the decisions involved in the aggressive exchanges of a wide variety of animals. Maynard Smith & Riechert's model contains some explicit suggestions about how escalation is achieved in fights between spiders, and it is interesting to compare these with ways the same ends are achieved in vertebrates. In particular, we consider warm-up or sensitisation effects and positive feedback from the opponent.

In the penultimate section, we discuss whether aggressive displays reveal information about the animal's motivational state to the opponent. As discussed in Chapter 9, the war of attrition model predicts that information indicating the likely display length will not be revealed to the opponent, but this prediction is in apparent conflict with the conventional ethological view that displays reflect an animal's motivational state.

The final question we consider concerns the mechanisms underlying the end of a fight. For the loser, this involves an apparently sudden change in the internal state, from one in which the aggressive tendency seems to predominate to one in which the fear tendency predominates to such an extent that the animal flees. How to represent this in motivational terms presents an unsolved problem and we discuss different approaches which have been brought to bear on it.

10.2 Benefits: the influence of resource value

The question of how changes in resource value affect motivation was discussed in Chapter 8 in relation to territorial and frustration-induced aggression. One general conclusion from this discussion was that resource value could affect aggressive behaviour in a number of different ways. When an animal builds up an attachment to a resource such as a spatial area or another conspecific, it is gradually assimilating into its internal representations what, in functional terms, represents an increase in the value of the resource, compared to the value for a would-be intruder or usurper. This process, mainly involving exposure learning, provides an example of functional rules being incorporated relatively indirectly into causal mechanisms (Chapter 1). In other cases, however, animals may be much more responsive to immediate variations in the value of a resource, and I illustrated this in Chapter 8 by considering territorial defence in nectar-feeding birds. A further example is provided by hermit crabs fighting over shells, where the intensity of fighting is higher when the shell is of good quality than when it is of poor quality (Dowds & Elwood, 1983). Similarly, peak fighting intensity in stags coincides with the time of the highest rate of conceptions (Clutton-Brock *et al.*, 1982). The level of escalation is also higher at higher resource values in fights between male spiders over access to a female (Riechert, 1979; Austad, 1983). In this case, resource value interacts with perceived cost to follow the V/K rule (Parker, 1974*b*) which can be used to determine the outcome, duration and degree of escalation (Austad, 1983).

This last example concerns the impact of the female's reproductive value on intermale aggression. It is interesting to note that the presence of a female facilitates male aggression in a wide variety of animals. Among arthropods, for example, male crayfish frequently fight in the breeding season, but this ceases when females are removed (Berrill & Arsenault, 1984). Similarly, in rodents such as deermice, housemice and rats, the presence of a female facilitates aggressiveness (Barnett, 1955; Barnett, Evans & Stoddart, 1958; Kuse & De Fries, 1976; Flannelly & Lore, 1977; Dewsbury, 1984b; Terman, 1984). There are several possible mechanisms which could underlie these influences, for example a specific sensory cue from the female could directly affect brain mechanisms controlling attack, or a socially mediated hormonal effect (such as an increase in testosterone levels) could act on neural structures to increase aggressiveness (cf. Flannelly & Lore, 1977).

These examples illustrate some of the ways in which the variable referred to as 'resource value' in game theory models may exert its influence in terms of proximate mechanisms; they range from the relatively indirect process of developing an attachment to a spatial area which provides a resource, to more immediate influences of changes in resource value on the readiness to fight and the intensity and persistence of fighting.

10.3   Costs: the motivational basis of RHP

Game theory models have alerted empirical researchers to the importance of RHP assessment for estimating in advance the likely costs of a fight. In terms of causal mechanisms, this will entail:

1 the recognition and assessment of the opponent's RHP;
2 a comparison with the animal's own RHP; and
3 a modification of behaviour in accordance with the outcome of this comparison.

Maynard Smith (1982) has argued that the following criteria are necessary for demonstrating that an RHP assessment strategy is operating. First, a difference between the contestants will be perceived by them and used to settle the contest without escalation. Secondly, behaviour occurring during the initial stages of the encounter will enable the contestants to perceive this difference. Thirdly, the particular signal used must be costly at a high value (if it were not, it would be vulnerable to bluff: see section 9.4). Finally, the signal should be correlated with fighting success.

Although functional models are not concerned with specifying which cues are used in the recognition and assessment of RHP, it is clear from the discussion in section 9.4 that weapons, body size and information about previous encounters with that opponent are the principal indicators of RHP.

There are many examples of animals assessing each other's fighting organs prior to a contest. These include open-mouthed threat in primates, which involves revealing the canines (Maynard Smith, 1982), the assessment of horn size in Stone's sheep (Geist, 1966), and assessment of the size of the major cheliped in hermit crabs (Neil, 1985). However, it should not be assumed that because an animal possesses fighting organs, they are necessarily used in assessment. Clutton-Brock *et al.* (1982) found no evidence that red deer used antler size in assessment of RHP.

In the previous chapter, we showed that body size can be used to indicate RHP in a wide range of animals (Table 9.1). Size may be assessed directly by visual cues, or indirectly through a feature which is related to body size, such as low-pitched sounds. One example of such an assessment has been investigated in toads (Davies & Halliday, 1978). Prior to mating, the male common toad clasps the female's back, behaviour which is referred to as amplexus. Males are carried around in this way for several days, during which time unpaired males try to dislodge them. Davies & Halliday found that all sizes of male had an equal chance of clasping a female initially, but by the time the female was spawning, larger males were more likely to be successful. By matching small, medium and large toads against one another as defenders or challengers, they showed that challengers of all sizes were more persistent against defenders which were smaller than them. Play-back experiments revealed that the toads used the depth of one another's croaks to assess size, since larger males make deeper croaks.

Clutton-Brock & Albon (1979) investigated whether red deer stags use one another's roaring performances to assess RHP. Holders of one-male social groups regularly engage in periods of reciprocal roaring with challengers. Differences between individuals and temporal changes in roaring rates were found to be correlated with fighting ability. Unlike the toads, which use croaks to assess size, stags roar only when there is no obvious size discrepancy between an individual and its rival. They are therefore using roaring to assess less obvious aspects of their condition. Unlike anatomical features, roaring varies throughout the course of the rut as stags lose condition, hence providing a sensitive indicator of fighting ability.

If, after a roaring contest, the approaching stag does not withdraw, the contest escalates further (Clutton-Brock *et al.*, 1982), leading to a parallel walk display, during which the stags invite contact by lowering their antlers. Even when physical contact begins, the protagonists engage in vigorous pushing, with their locked antlers, ending when one of them is pushed rapidly backwards, breaks contact and runs. Such trials of strength can themselves be viewed as a form of RHP assessment since they do not reflect the ultimate potential of the animals to damage one another. In fact, there are many similar examples where animals fight by testing one another's strength and stamina, rather than risking a more damaging encounter. D. C. Blanchard *et al.* (1978) reported that fights between equally matched brown bears were of this type. Ramming displays, which appear to enable contestants to assess one another's strength and stamina, have been observed in males of the fish *Haplochromis burtoni* (Mosler, 1985). Similarly, mouth-wrestling has been found in the cichlid *Nannacara anomala* (Jakobsson *et al.*, 1979). In male bowl and doily spiders, jawlock displays are found (Suter & Keiley, 1984), and these last longer when the opponents are similar in size.

Whether weapons, size or tests of strength form the basis of RHP assessment, a comparison of the opponent's RHP cues with those indicating the animal's own fighting condition must be made. Such a comparison would be strongly influenced by the cumulative effect of past fighting experiences. The basis of the comparison would be established early in social development. In Chapter 7, we considered the function of rough and tumble play and suggested that it may promote flexibility in social responses, including the learning of assessment strategies. In Chapter 8, we considered studies of mice which showed that adult aggressiveness was lower in males reared with other males than those reared alone or with a group of females, and it was suggested that males generally learned assessment strategies through their low-intensity juvenile fights (Poole & Fish, 1975). Bekoff (1981), discussing the early fights of coyotes, suggested that they learn behavioural controls very rapidly during the course of these relatively severe fights (see also Bekoff *et al.*, 1981). Wilson & Franchlin (1985) found that male guanacos, *Lama guanicoe* (a gregarious ungulate), spend 3 to 4 years in all-male groups where they learn to assess size and fighting ability by means of chest rams and play fighting.

During adulthood, an animal's own RHP may be subject to abrupt changes, for example following illness or injury (Clutton-Brock *et al.*, 1982), and to more gradual declines during ageing, or periods of weight loss. In this context, it is interesting to note that Bevan *et al.* (1960) found that the decline in readiness to attack by male mice following castration was related to their degree of weight loss (Chapter 8).

Whatever the long-term influences on the assessment process, the immediate comparison must be between some 'reference value' (R), indicating the animal's estimate of its own RHP, and its perception of the opponent's RHP (O). Each animal must therefore possess such a reference value before entering a conflict (and it may be temporarily enhanced by the presence of an ally or allies: section 9.4). One very simple rule would be for a value of $R/O > 1$ to increase the likelihood of attack in proportion to its numerical size, and a value of $R/O < 1$ to decrease it. A further refinement to this rule can be derived from the model of Parker & Rubenstein (1981), discussed in section 9.6, which considered interactions between RHP and resource value. Where resource value is unequal, the ratio R/O can be replaced by the assessor value, i.e. $(Va/Ka)/(Vb/Kb)$, Va being the animal's own resource value, and Vb the value to its opponent; Ka and Kb represent the costs of fighting for the animal and its opponent, which will reflect their respective RHPs.

In the control system model of territorial aggression outlined in Chapter 8 (Fig. 8.2), assessments of the cost of fighting (i.e. RHP assessment) were shown to affect the decision process which determines the balance between the tendencies for aggression and fear. In experimental studies using inanimate objects, larger-sized objects tend to evoke fear responses instead of, or as well as, attack (Schaller & Emlen, 1962, Hoffman *et al.*, 1974, for the domestic fowl; R. J. Blanchard, Mast & D. C. Blanchard, 1975, for rats). We have already noted that smaller conspecific opponents are more likely to be attacked, whereas larger-sized ones tend to evoke fleeing. Experimental studies have shown that female mice, which would not normally attack males near to their own size, will more readily attack smaller opponents (Edwards, 1969; White, Mayo & Edwards, 1969). It seems likely, therefore, that perceived RHP superiority affects the balance of aggressive and fear tendencies in favour of aggression, rather than simply increasing the likelihood of attack, as indicated above.

10.4 A two-factor model of aggressive behaviour in funnel-web spiders

For the purposes of the discussion in the previous section, it was assumed that only two outcomes were possible from the interaction

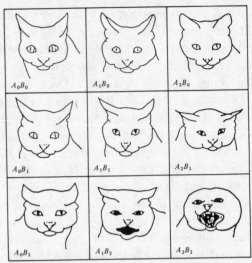

Fig. 10.1 Facial expressions shown by domestic cats under the influence of aggressive (A) and fear (B) tendencies. A_0B_0 shows a low level of both: moving to the right, A increases, moving downwards B increases, so that A_2B_0 is the highest level of A/B, and A_0B_2 the highest level of B/A. (Modified from Leyhausen (1956), p. 137, and (1979), p. 195.)

between aggression and fear tendencies, either attack or fear behaviour according to which tendency was highest. This is an oversimplification which neglects the wide range of threat postures which have been described in ethological studies of fish, birds and mammals. In many cases, there is evidence that these can be explained in terms of conflicting tendencies to attack or to flee, so that there will be a range of possible responses from outright fear to outright attack, with intermediate displays reflecting different motivational levels of aggressive and fear tendencies (Hinde, 1970; Baerends, 1975). Figure 10.1 shows Leyhausen's (1956, 1979) diagram of the facial expressions of the domestic cat under the influence of different levels of aggression and fear. Figure 10.2 illustrates in more general terms how several possible outcomes can result from the competing tendencies for aggression and fear. This is sometimes referred to as the 'two-factor' model of displays.

The motivational model of the agonistic behaviour of the funnel-web spider proposed by Maynard Smith & Riechert (1984) begins with the assumption that conflicting tendencies for aggression and fear produce a variety of outcomes according to their relative levels. How this applies

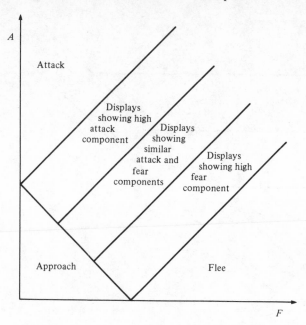

Fig. 10.2. Graphical representation of the interaction between the tendencies for aggression (A) and fear (F), to produce a series of displays differing in motivational state. (Reproduced with permission of F. A. Huntingford.)

to the particular case of the funnel-web spider is shown in Fig. 10.3. Here I have used Maynard Smith & Riechert's own notation, Y depicting the fear tendency and X the aggressive tendency (instead of F and A as before). When the level of Y greatly exceeds that for X, the animal withdraws permanently (lower right). When the difference is smaller, temporary retreat occurs. If the two tendencies are approximately equal or if X exceeds Y, the contest continues, with a level of aggressive acts being determined by the absolute level of X. (Here there is a departure from the classical ethological view that it is the relative levels of aggression and fear that determine the outcome, as shown in Fig. 10.2.)

Maynard Smith & Riechert's model is based on the results of Riechert's studies of agonistic encounters in the funnel-web spider (referred to in the previous chapter). It differs from the control system models described earlier in this book in several respects. First, it begins with variables which have been identified in functional game theory models (such as RHP assessment and resource value), rather than with those which might be of immediate causal significance. Secondly, it

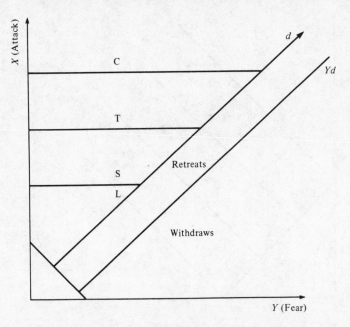

Fig. 10.3. Graphical representation of the interaction between the tendencies for attack (X) and fear (Y) in the funnel-web spider, to show their interaction according to the model of Maynard Smith & Riechert (1984).

d is the hypothetical dominance boundary where X and Y are approximately equal ($Z = 0$). Below the level represented by the line Yd, where $Z(X - Y)$ is large and negative (< -5), permanent withdrawal occurs. When Z is small in value, and negative (0 to -5), i.e. between lines d and Yd, temporary retreat occurs. If Z is positive, the animal performs aggressive acts, in the following increasing order of escalation: L (locate), S (signal), T (threat), and C (contact), according to the absolute level of X.

depicts the interaction between two animals, rather than motivational processes within a single animal. Thirdly, it incorporates numerical values for the various parameters so that simulated and real aggressive exchanges can be compared. The predictive value of the model for features such as the eventual winner and the degree of escalation under different conditions (such as resource value, relative sizes of the contestants, and population line) has proved reasonably accurate.

As shown in Fig. 10.3, the spiders' agonistic behaviour can be classified into four categories of increasing intensity: locate, signal, threat and contact. A fight consists of a number of bouts, each ending in retreat by one of the contestants, and the series of bouts ultimately ends when one of the participants withdraws permanently (Fig. 10.3). Within each bout, both spiders tend to escalate, and this occurs more rapidly when a spider has a weight advantage over its opponent. In one of the two populations studied (from a desert grassland habitat), variations in web-site value were important, whereas in the other (from a desert riparian area), they were not. Where they were important, only the behaviour of the owner was affected by the resource value. In the desert environment, web-sites were generally a less valued resource than in the grassland, and the associated contests tended to be shorter and less costly.

Maynard Smith & Riechert divided aggressive and fear tendencies into temporary components, which rise during a bout, and permanent ones, which change in relation to information obtained about relative fighting ability (RHP) of the opponent. Overall aggressive and fear tendencies are the sum of the two respective components.

They presented their model as a flow diagram in which the interaction is seen as a series of moves, with the two opponents acting alternately, so that at each instant there is a mover (M) and a non-mover (non-M) (Fig. 10.4). The input of the flow diagram consists of a number of variables (1, Fig. 10.4) which affect the values of X and Y in the mover (2–4, Fig. 10.4). The difference between X and Y (Z) is used to determine the following: first, whether the mover permanently withdraws or not (according to rules which incorporate the value of the site to the owner); secondly, whether it retreats; and thirdly, the type of aggressive act it will use if it fights (5–7, Fig. 10.4). The behaviour of the mover will affect the motivational state of the non-mover by providing information about their relative weights (8, Fig. 10.4) and by producing a temporary (8, Fig. 10.4) and permanent (9, Fig. 10.4) increase in its aggressive tendency. After a bout, the next one is started by the spider with the higher Z value (10, Fig. 10.4).

The model was tested against field and laboratory data from interactions involving three weight categories from the riparian population (<10 per cent difference, owner +10 per cent > intruder, or +10 per cent < intruder), and from the grassland population, the latter at three classes of web sites (excellent, poor and average). Data were also obtained from interactions between two weight categories (<10 per cent difference or >10 per cent difference) for hybrids, and (in the laboratory) interactions between feral riparian and grassland spiders of various weight categories. The energetic costs of each contest were calculated in terms of estimated energy expenditure, based on the types of acts involved (i.e. the degree of escalation).

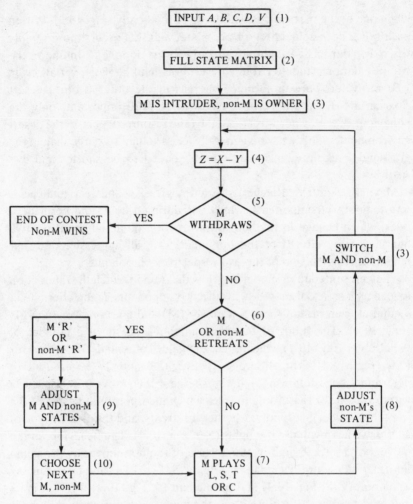

Fig. 10.4. Flow diagram for simulations of spider agonistic interactions. Input variables: A = rate of increase of aggression; B = rate of increase of fear; C = effect of ownership; D = effect of relative body weight; and V = effect of territory quality on aggression and fear. X = level of aggression. Y = level of fear; L = locate; S = signal; T = threat; C = contact; 'R' = retreats. (From Maynard Smith & Riechert (1984), p. 568, original caption.)

The model reproduced the overall features of real contests, for example that they entailed bouts which consisted of behavioural acts occurring as an escalated sequence. Moreover, it also reproduced most of their detailed characteristics: higher costs were incurred in fights over excellent sites when the intruder was larger than the resident (section 9.6); grassland spiders showed more costly fights than riparian ones; and contests were won by the larger spider if there was a pronounced size difference, and usually by the owner if not.

Maynard Smith & Riechert's model represents a considerable achievement in producing a working motivational model based on functional considerations, empirical data and the ethological view of competing tendencies as an explanation of agonistic displays. Simulations of encounters using the model have also managed to mimic the patterns of act transitions in real fights and to provide a simple and plausible explanation for genetic differences between the riparian, grassland and hybrid populations (the latter showing a greater probability of escalating than either of the parental lines).

Despite these considerable achievements, the model has some restrictions as a motivational model. It is not concerned with the exact nature of the mechanisms involved. For example, it begs the question of *how V* (resource value) affects the owner of a web, whether by exposure learning (as suggested earlier for the value of a territory), or by shorter-term adjustments equivalent to the learning mechanisms in nectar-feeding birds (Chapter 8), or by some other alternative means.

10.5 Mechanisms underlying escalation

At various places in Chapter 9, I referred to the notion of escalation. This entails the sequential replacement of low-cost, low-risk acts occurring early in the contest by those which are energetically more costly and entail a higher risk of injury, as the contest progresses. As indicated above, Riechert's studies of the funnel-web spider revealed a graded series of aggressive behaviour from L (locate) to C (contact), each representing energetically more costly acts. Studies of a variety of other animals have also shown that energetically more costly acts tend to occur as a fight progresses (e.g. in two fish species: Dow, Ewing & Sutherland, 1976, Jakobsson *et al.*, 1979; in iguanas: Rand & Rand, 1976). Higher-cost actions are more likely to bring the animals into closer proximity, thus involving a greater risk of injury and a greater likelihood of being followed by damaging actions, such as biting.

In their model, Maynard Smith & Riechert assumed that the level of aggressive tendency (X) determined the degree of escalation, independently of that for fear (Y). The level of X depended on a genetic variable (A) controlling the rate at which it builds up within a bout, and on variables representing the permanent aggressiveness of the owner (C), resource value (V) and size difference (D). A generates a gradual increase in the temporary level of X during bouts (Xt). There will also be a permanent increase in aggression (Xp) as a function of the reaction to the opponent's behaviour, and this is largely determined by an assessment of the weight difference (D) and a constant (G), which increases with the level of the opponent's act on the escalation scale. The influence of D and G is seen as multiplicative, i.e. a large weight difference and a higher-intensity activity in the opponent combine to increase the degree of escalation.

Ultimately, the degree of escalation is the sum Xt and Xp. Escalation is therefore viewed as depending on two variables, one of which (Xt) is internally generated and depends on how long the bout has progressed, and the other of which (Xp) is dependent on the opponent's behaviour and an assessment of the size difference.

The mechanisms underlying escalation in vertebrates can also be understood in terms of a combination of internal and external processes. Internal processes would be represented by 'warm-up' or sensitisation effects, which have been studied in fish and rodents (e.g. Peeke, 1982; Potegal & tenBrink, 1984): the rate and level of attack begins at a low level and increases as the fight progresses. Peeke (1982) has suggested that both stimulus-specific and motivation-specific components are involved in generating sensitisation effects. The first occurs as a consequence of the continued presence of a specific opponent, and the second as a consequence of the time for which the motivational system had been activated. Such mechanisms can provide a partial explanation of escalation in terms of internal processes but they take little account of the dynamics of an encounter and functionally relevant situational variables, such as an assessment of the opponent's RHP, and resource value.

Dow *et al.* (1976) studied aggressive encounters between pairs of the fish *Amphyosemion striatum*, matched for RHP. Individual aggressive acts could be arranged as an escalating sequence of increasing intensity. There was a sequential replacement by higher-level acts as the fights progressed, until the losing fish suddenly signalled its capitulation by 'fin clamp'. Dow *et al.* suggested that a mutual positive feedback operates as the fight progresses, with each fish making or responding to a series of 'bids' from its opponent, until one of them reaches a point at which it is

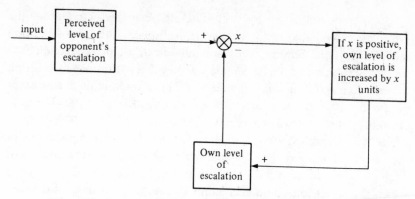

Fig. 10.5. Discrepancy-activated mechanism to depict matching of level of escalation to that of opponent, as suggested by Dow *et al.* (1976).

not able to proceed to a higher level (see section 10.7). This is shown diagrammatically as a discrepancy-activated process in Fig. 10.5.

If we consider both sensitisation and externally produced positive feedback together, it is likely that sensitisation occurs at a slightly different rate in one fish than the other. In the absence of external influences, one fish would tend to escalate at a slightly slower rate than the other. In matching the bids of its opponent, the slower fish would need to compensate by boosting its level of escalation so as to keep up with that of its opponent (Fig. 10.5). Whether this has any significance for which animal wins the encounter, i.e. whether the animal which leads the bidding is more likely to win, is uncertain. (We discuss the question of how fights end in section 10.7.)

10.6 Communication in aggressive displays

The conventional ethological view of threat displays is that they reflect the animal's underlying motivational state, each display occurring at a particular motivational mixture of aggressive and fear tendencies (Fig. 10.2). Given this relationship, an opponent would have access to some information about the animal's motivational state, and therefore its likelihood of particular actions in the future. There is also agreement within the ethological tradition that during the course of evolution, displays have been subject to changes concerned with their signalling function, variously described as ritualisation (Blest, 1961). A display may, for example, have become fixed at a 'typical intensity' (Morris, 1957), i.e. a fixed level of display is given regardless of the underlying motivational state.

In apparent conflict with the ethological view which emphasises the information available to the opponent in displays, the war of attrition model (Maynard Smith, 1974) and modifications of it (section 9.5) predict that animal displays should have evolved so as not to transmit information about fighting intentions. These apparently inconsistent predictions have been discussed by Dawkins & Krebs (1978), Krebs & Dawkins (1984), by Maynard Smith (1982) and by Caryl (1979, 1981).

Krebs & Dawkins (1984) concentrated on the broad implications of the game theory approach to animal signals. They argued that instead of evolving for effective information transfer, animal signals provide a means by which one animal manipulates another. Taking this view, emotional expressions such as threat displays can be seen as extensions of the way animals will manipulate any object in the environent to their advantage. Since another animal is more complex than an inanimate object, to ensure effective manipulation, its actions have to be interpreted correctly. Krebs & Dawkins refer to this as 'mind-reading', and they view animal signals as a result of the co-evolution of manipulation and 'mind-reading'.

Maynard Smith (1982) was more concerned with assessing the evidence for the war of attrition model than with such broader implications. His original model predicted that, when no asymmetries are involved, contests should be variable in length, and display intensity should be constant irrespective of how much longer the animal will continue, therefore giving no information about future intentions to the opponent. In support of this prediction, he cited several studies which seemed to show that the eventual winner or loser cannot readily be distinguished until near the end of a contest. One was Simpson's (1968) detailed analysis of the display components of the Siamese fighting fish, *Betta splendens*. He found no differences between the behaviour of the eventual winner and loser, at least until near the end of the contest when the proportion of time the gill covers were raised differed. Another study was that of Dow *et al.* (1976), on the fish *Aphyosemion striatum*, described in the previous section. They found that the level of aggression was high in both fish until near the end of the encounter when 'fin clamp' from the losing fish signalled the (abrupt) end of the fight. Similarly, Jakobsson *et al.* (1979) studied encounters between males of the cichlid *Nannacara anomala* matched for size, residency and dominance, and found a great similarity between the behaviour of the eventual winners and the losers during the early and middle parts of the fight. There were, however, some differences, in particular a greater tendency by eventual winners to retaliate when bitten.

In the section on the war of attrition model (9.5), we indicated that Caryl (1979) had recast the original equation in terms of costs rather than duration (following Norman *et al.*, 1977), thus predicting that threat displays should not transmit information about the cost an individual is prepared to pay in a particular contest. Caryl then reanalysed the empirical evidence from four previously published studies of bird displays, to examine this prediction. In all four studies, the display shown in an encounter could be related to whether the bird subsequently attacked, stayed or escaped. Caryl found that although the ability to predict escape was good, attack could not readily be predicted from the type of display. He also showed that the display which most strongly predicted attack did not have a marked effect on the other bird's behaviour. These findings suggest that there is considerable emancipation of the aggressive displays from their underlying motivation and they were taken to support in broad terms the prediction of the war of attrition models rather than the classical ethological view. It is interesting to note that the displays did predict information about escape, since, as Maynard Smith (1982) pointed out, capitulation is the one intention which it is advantageous to signal, therefore avoiding the possibility of further attack.

Hinde (1981) argued that Caryl (and also Dawkins & Krebs, 1978) had overemphasised the differences between ethological and game theory approaches to agonistic displays. For example, as indicated above, the ethological literature emphasises that many displays become ritualised during the course of evolution, leading to a typical intensity and a degree of emancipation from the underlying motivational state. Hinde's comments on the finding that display components do not accurately predict attack highlight the difference in approach between game theory and the analysis of displays as social interactions. He viewed the lack of predictability in displays as a consequence of the actor being uncertain about its next action because this depends partly upon the response of the reactor. In this way, Hinde emphasised the changing and interactive quality of the interactions in an agonistic sequence.

The modified war of attrition model predicts only that the maximum cost an animal will pay in a contest is not revealed to its opponent. Hinde is surely correct, therefore, to identify a gap between this very broad prediction and the question of whether winners and losers can be identified from details of the social exchange during the early and middle parts of a fight. Commenting on the fish studies cited by Maynard Smith (see above), Turner & Huntingford (1986) make a similar point, remarking that 'it seems strange to regard an individual as "intending" to win or lose at the start of a contest' (p. 968). Of course, future winners and losers may be readily identifiable where there are RHP differences between them, and we should note that the war of attrition predictions apply only to cases where displays relating to RHP assessment have occurred and failed to reveal any differences.

Maynard Smith (1982) stressed the importance, from the game theory perspective, of considering RHP signals separately from those concerning intentions (section 9.4). In theory, RHP communication, which involves variations in a single type of behaviour such as roaring or pushing, can be distinguished from communication about internal state, which involves selection of one act from a number of possible forms of behaviour (Enquist, 1985; Turner & Huntingford, 1986). But it may be difficult in practice to separate the two forms of behaviour if they interact and influence one another during the course of a conflict.

In the case of Siamese fighting fish, where a detailed analysis of matched pairs revealed no differences between the display components of eventual winners and losers (Simpson, 1968), more recent evidence indicates that an assessment stage precedes a more prolonged and persistent escalation stage (Bronstein, 1981, 1983). Even in matched pairs of fish, the extent of attack early in the testing period was found to correlate positively with the subsequent durations of attack (Bronstein, 1985a, b), suggesting that differences in levels of aggressiveness between the fish were apparent early on. It was also found that winners of paired encounters were typically those which had built larger nests (Bronstein, 1985b), suggesting that internal motivational state may reflect the higher resource value in this case. These studies highlight the practical difficulty of ensuring that no asymmetries are operating, and hence of demonstrating that the conditions under which the war of attrition model applies are operating.

In an earlier section (10.3), we discussed the way in which the results of RHP assessment would influence subsequent motivation, by altering the balance of aggressive and fear tendencies. The outcome will presumably affect subsequent fighting intentions. Whether RHP is communicated to the opponent once the contest is underway is not clear from game theory models. Although both Parker's (1974b) RHP model and the war of attrition model (Maynard Smith, 1974) assume that information is available before the animal enters the aggressive encounter, subsequent models such as those of Parker & Rubenstein (1981) and Enquist & Leimar (1983) have emphasised the gradual acquisition of information during a fight. In motivational terms, it would certainly seem unrealistic to regard an animal as being able to decide in advance the detailed course of its aggressive behaviour, particularly if RHP information becomes available only as a consequence of interacting with the contestant.

The war of attrition model predicts that the maximum cost each animal is prepared to pay will be set in advance, and will not be revealed to the opponent. Since the cost an animal is prepared to pay in a particular contest will depend on the value of the resource, in those cases where one or both animals do not possess information about the resource value at the beginning of a contest (Chapter 9), the cost they are prepared to pay may change as further information becomes available. In other cases, resource value itself may vary as the fight progresses, again altering the maximum cost.

These considerations indicate that the war of attrition model will apply to only a restricted range of conditions. However, if we assume that the conditions of the model are met, its prediction does have implications for a motivational analysis. One possibility is that the motivational balance (level of A/F) at which the animal will capitulate is set in advance, yet exerts little or no influence on the exact nature of the fighting sequence until just prior to giving up. This corresponds to the analogy of a series of bids used by Dow *et al.* in their description of the fighting sequence of two fish (see section 10.5), and also by Maynard Smith (1972) in his first article on the game theory approach (i.e. before the war of attrition model). Maynard Smith's initial argument was as follows. During an aggressive encounter, each animal is under competing motivation to continue attacking or to retreat, the precise strength of each depending on a combination of internal and external causal factors. However, it will not pay either contestant to reveal the strength of this motivation (since it could be used to predict the giving-up point), in an analagous way that it would not pay anyone to reveal the level at which they would settle when negotiating, or indeed whether they were near to that level. Natural selection, Maynard Smith argued, would favour a sharp switch at a threshold level of motivation. As long as the motivational balance still favours attack, it should occur at full intensity, but once the threshold level is reached, there should be a sudden change to escape behaviour.

The implications of this description for the process of escalation, discussed in section 10.5, are that it can only be a one-way process. There will be a gradual build-up from lower- to higher-level activities, both in terms of cost and intensity, from the beginning to the end of the fight but any movement from higher- to lower-level activities can occur only when the giving-up point is reached. Thus, changes in motivational state in the direction of increased fear tendency will not be shown throughout the fight, but will be revealed only at the point of capitulation. There must therefore be a mechanism on the output side which holds the degree of escalation at the level which has been reached so far, despite any changes in motivational balance, unless the critical value has been reached: at this point, the 'hold' mechanism will be overridden and the true motivational balance will control the output. This motivational description fulfils the predictions of the war of attrition model, yet the conflict is still viewed as an interacting sequence which depends on the mutual responses of actor and reactor.

10.7 How do fights end?

In the previous two sections, I have referred to the gradual escalation of a fight to a point at which one animal ceases attacking and flees. Where unequal RHP is involved, I suggested (section 10.3) that the level of F/A would increase in the animal with the lower perceived RHP. It is relatively easy to see how the balance of fear relative to aggression could, under these conditions, readily exceed a threshold value at which fleeing occurred. A similar rapid switch also occurs when this threshold is exceeded during the course of a more evenly matched encounter. However, as argued in the previous section, as long as the motivational balance favours attack, it should be maintained at full intensity; once the balance favours fear, there should be a rapid and irreversible switch to escape behaviour.

Sudden transitions from an apparently high level of attack to fleeing have been observed in many cases of animal conflict. Zeeman (1976) has applied catastrophe theory, a form of mathematics developed to explain systems which show sharp transitions, to these motivational transitions. As Toates (1980) noted, catastrophe theory can be viewed as an extension of the two-dimensional representation of the interaction of aggression and fear tendencies (Fig. 10.2). This is shown in Fig. 10.6 (from Toates, 1980). The distinctive feature of catastrophe theory is that it illustrates, in three-dimensional space, how a small increase in the level of fear tendency relative to that for attack, can produce a rapid and extensive move on the cusped surface representing behaviour (Fig. 10.7). Owing to the nature of this cusped surface, the reverse change, i.e. a small increase in the level of attack relative to that for fear, would not produce a large change on the behaviour surface.

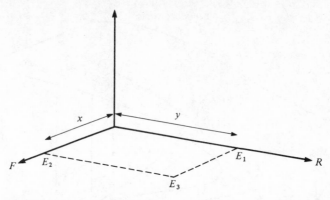

Fig. 10.6. Catastrophe theory representation of causal factors. $R =$ rage dimension, $F =$ fear dimension. E_3 therefore has both a rage component E_1 (magnitude y) and a fear component E_2 (magnitude x). (From Toates (1980), p. 214, original caption.)

There are, however, a number of difficulties concerning the application of catastrophe theory to the interaction of aggression and fear in this way. The first is that such a graphical description may do little more than provide a pictorial analogy of what is happening in the animal, and may detract from the search for the mechanisms underlying the sudden transition from attack to escape shown by a losing animal. Applications of catastrophe theory to the biological and social sciences have been criticised for their exaggerated claims, incorrect reasoning and far-fetched assumptions by Zahler & Sussmann (1977), who concluded that catastrophe theory provides no advantage over better-established mathematical tools. Similarly, Croll (1976) has criticised applications of catastrophe theory, particularly to the social sciences, concluding that identification of the correct catastrophe graph in any given case requires precise quantitative information, which is lacking for most behavioural examples, including Zeeman's (1976) analysis of aggression. There is, therefore, no way of telling whether the catastrophe graph shown in Fig. 10.7 is the correct one. Nevertheless, there is still a need for mathematical representations of non-linear problems which can be applied to behaviour.

Perhaps a more fruitful approach to analysing the discontinuity in behaviour at the end of a fight would be from a motivational viewpoint. In general terms, the problem concerns which of two motivational tendencies, incompatible at a motor level, attains behavioural dominance. The sequential replacement of one type of behaviour (A) by another (B) can often be explained by a rise in the causal factors for B

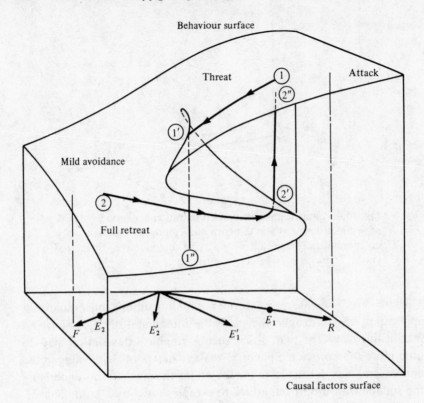

Fig. 10.7. Zeeman's catastrophe theory model of aggression in dogs. The model provides a three-dimensional extension of the causal factors shown in Fig. 10.6, using the same abbreviations. While the causal factors surface is flat, the corresponding behaviour surface contains a folded cusp, so that in this area each combination of causal factors has two outcomes. An angry dog made more fearful is shown by the line (1): its behaviour becomes gradually more influenced by fear until the point (1′) where it shows a sudden decrease to (1″), i.e. a sudden change to full retreat. In contrast, a fearful dog which is angered is shown by the line (2): its behaviour shows a steady level of fear behaviour, until it reaches point (2′), when it suddenly attacks (2″) represented by jumping to (2″). (From Toates (1980), p. 216, based on Zeeman (1976).)

(a)

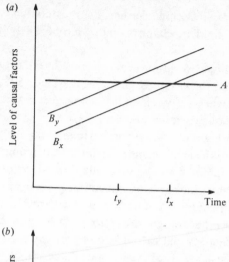

(b)

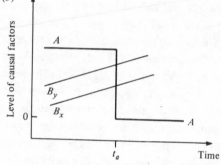

Fig. 10.8. Level of causal factors as a function of time. In (a), behaviour *B* ousts activity *A* by competition. In (b), *B* is disinhibited by *A*. (From McFarland (1969), p. 238, original caption, slightly abbreviated.)

(McFarland, 1969; Fig. 10.8a), referred to as competition, and presumably this can account for the replacement of lower-level by higher-level activities during the course of escalation. However, when a fight suddenly comes to an end, disinhibition, the alternative mechanism suggested by McFarland (1969), would appear to be involved (Fig. 10.8b). Disinhibition usually occurs when it is important to interrupt periodically the dominant activity, or when it is important to complete a particular sequence of activities and to inhibit others until this is achieved (McFarland, 1985). Abrupt cessation of fighting probably falls into the second category, since it is important to maintain fighting at full intensity until the switch-over point is reached. The influence of the competing fear tendency remains inhibited by the aggressive tendency until this point, when the level of causal factors for the aggressive tendency apparently shows a rapid decline. Earlier, I suggested that this apparent decline was due to a 'hold' mechanism on the output side being overridden.

Although McFarland's original two graphs can be considered as two examples from a wider variety of cases (Roper & Crossland, 1982), disinhibition is the only one where a precipitous decline occurs in the level of causal factors for one of two activities.

The crucial question of what determines the switching-point still remains. Dow *et al.* referred to the sudden end of fighting in one animal as due to its 'overloading'. In their study, subsequent winners were more willing to escalate in later stages of the encounter. Jakobsson *et al.* reported that eventual winners were more likely to retaliate when bitten, and that this difference became more pronounced towards the end of the encounter. This suggests that the end-point is not sudden or unexpected, for the winner at least. Earlier, we speculated about the influence of RHP assessment on escalation of a fight. Extending this, one possibility is that as the fight progresses, one animal obtains information which tends to confirm that its own RHP value is greater than that of its opponent. Accordingly, the opponent perceives that its own RHP value is smaller. For the former, the disparity acts to facilitate the tendency to attack and escalate the sequence, whereas for the latter it acts as an inhibiting factor, since it tends to increase the fear tendency relative to that for aggression. Although, as we have argued, the output will still favour aggression, there comes a point when a critical value for F/A is reached, and a switch to escape occurs at this point.

10.8 Conclusions and suggestions for future research

This chapter considered the mechanisms underlying the functional variables which game theory models identified as important in animal conflicts. We considered mechanisms whereby RHP influenced motivation, suggesting that perceived RHP superiority increases the balance of aggression and fear tendencies in favour of aggression.

Maynard Smith & Riechert's motivational model of aggression in spiders was based on a combination of the ethological two-factor view of animal displays, functional variables from game theory models and Riechert's empirical results. It accurately simulated the patterns of act transitions in real fights and it also portrayed the influence of resource value, size and population characteristics on aggressive behaviour. In this model, escalation depended on two variables, one generated internally and the other generated externally. Similarly, internal and external processes contribute to escalation in vertebrates; external processes take the form of attempts to match the level of attack to that shown by the opponent if this is at a higher level.

The war of attrition model has been regarded as inconsistent with the ethological view that motivational state is revealed in agonistic displays. We concluded that the conditions under which the war of attrition model applies are narrow ones, and that when these conditions are fulfilled, the model predicts only that an animal will not reveal the maximum cost it is prepared to pay in fighting its opponent. This implies that as long as the motivational balance (between aggressive and fear tendencies) has not reached a threshold level (determined by the maximum cost function), attack should occur at full strength. Once the threshold level is reached, a sudden change to escape will occur.

It is clear from the studies reviewed in this chapter and the previous one that the functional approach, as represented by game theory models, has been widely applied to fighting strategies but that motivational issues have seldom been considered in conjunction with it. Of course, Maynard Smith & Riechert's model of spider aggression is the outstanding exception. One general aim in future research could be to provide specific working models, derived from functional variables, of other cases of well-researched aggressive interactions, for example in the fish species used for studies of sensitisation (section 10.5). Specific models can be used to make clear tests of the motivational implications of game theory with respect to independent variables such as RHP, ownership and resource value, and dependent variables such as escalation and the prediction of the outcome.

There are many existing ethological analyses of the sequences of aggressive interactions. Where relevant, these were referred to in this chapter. However, many potentially useful older studies cannot readily be applied to predictions from game theory models, since the relevant variables were not measured. Rasa (1969), for example, found that a low index of fearful behaviour at the beginning of a fight was correlated with subsequent victory in conflicts between pairs of juvenile pomacentrid fish. This potentially interesting finding is, however, of limited value for predictions derived from game theory because no information was given on RHP difference (relative size or past experience) between the fish. More generally, a problem in applying game theory predictions to real fights lies in the quite different predictions involved when there are RHP differences and when there are not. Often, these two cases are clearly distinguished in the models but they may not be so readily distinguished in real fights: it is difficult in practice to ensure that animals are not responding to subtle RHP cues. One approach which has been used in research derived from game theory models is to compare fights between different weight categories, thus enabling a comparison of the effects of different magnitudes of RHP difference to be undertaken.

Finally, one outstanding problem concerning the mechanisms underlying fighting sequences is to provide an account of the motivational changes occurring when the loser capitulates. The account we have given in this chapter provides a number of hypotheses which could be investigated. First, it was suggested that escalation is a one-way process, i.e. the level of attack is maintained at the previous highest level even if the A/F balance alters in the direction of F. It was suggested that the threshold, or giving-up point, is set in advance but not revealed. However, it should be possible to manipulate this point by altering the resource value, on the assumption that the cost the animal is prepared to pay will be higher if the potential benefit is higher.

Further reading

Caryl, P. G. (1979). Communication by agonistic displays: what can game theory contribute to ethology? *Behaviour*, **68**, 136–69.

Davies, N. B. & Halliday, T. M. (1978). Deep croaks and fighting assessment in toads *Bufo bufo*. *Nature*, **174**, 683–5.

Hinde, R. A. (1981). Animal signals: ethological and games-theory approaches are not incompatible. *Animal Behaviour*, **29**, 535–42.

Maynard Smith, J. & Riechert, S. E. (1984). A conflicting-tendency model of spider agonistic behaviour: hybrid–pure population line comparisons. *Animal Behaviour*, **32**, 564–78.

REFERENCES

Adams, D. B. (1979). Brain mechanisms for offense, defense, and submission. *The Behavioral and Brain Sciences*, **2**, 201–41.

Adams, D. B. (1980). Motivational systems of agonistic behavior in muroid rodents: a comparative review and neural model. *Aggressive Behavior*, **6**, 295–346.

Ahlen, I. & Andersson, A. (1970). Breeding ecology of an eider population on Spitsbergen. *Ornis Scandanavia*, **1**, 83–106.

Alados, C. L. (1986). Aggressive behaviour, sexual strategies and their relation to age in male Spanish ibex (*Capra pyrenaica*). *Behavioural Processes*, **12**, 145–58.

Alberts, J. R. & Galef, B. G. (1973). Olfactory cues and movement: stimuli mediating intraspecific aggression in the wild Norway rat. *Journal of Comparative and Physiological Psychology*, **85**, 233–42.

Alder, E. M., Godfrey, J., McGill, T. E. & Watt, K. R. (1981). The contributions of genotype and sex to variation in mating behaviour between geographical subspecies of the bank vole (*Clethrionomys glareolus* Schreber). *Animal Behaviour*, **29**, 942–52.

Allee, W. C. (1942). Social dominance and subordination among vertebrates. *Biological Symposia*, **8**, 139–62.

Allee, W. C., Collias, N. E. & Lutherman, C. Z. (1939). Modification of the social order in flocks of hens by injection of testosterone propionate. *Physiological Zoology*, **12**, 412–40.

Allee, W. C., Foreman, D., Banks, E. M. & Holabird, C. H. (1955). Effects of androgen on dominance and subordination in six common breeds of *Gallus gallus*. *Physiological Zoology*, **28**, 89–115.

Amsel, A. & Roussel, J. (1952). Motivational properties of frustration. I. Effect on a running response of the addition of frustration to the motivational complex. *Journal of Experimental Psychology*, **43**, 363–8.

Andersson, M., Wiklund, C. G. & Rundgren, H. (1980). Parental defence of offspring: a model and an example. *Animal Behaviour*, **28**, 536–42.

Andrew, R. J. (1969). The effects of testosterone on avian vocalisations. In *Bird Vocalisations*, ed. R. A. Hinde, pp. 97–130. Cambridge: Cambridge University Press.

Andrew, R. J. (1972). Recognition processes and behavior with special reference to effects of testosterone on persistence. In *Advances in the Study of Behaviour*, vol. 4, ed. D. S. Lehrman, R. A. Hinde & E. Shaw, pp. 175–208. New York & London: Academic Press.

Andrew, R. J. (1980). The functional organization of phases of memory consolidation. In *Advances in the Study of Behavior*, vol. 11, ed. R. A. Hinde, C. Beer & M. C. Busnel, pp. 337–67. New York & London: Academic Press.

Andrew, R. J. (1983). Specific short-latency effects of oestradiol and testosterone on distractability and memory formation in the young domestic chick. In *Hormones and Behaviour in Higher Vertebrates*, ed. J. Balthazart, E. Prove & R. Gilles, pp. 463–73. Berlin: Springer-Verlag.

Andrew, R. J. & Rogers, L. (1972). Testosterone, search behaviour, and persistence. *Nature*, **237**, 343–6.

Archer, J. (1970). Effects of population density on behaviour in rodents. In *Social Behaviour in Birds and Mammals*, ed. J. H. Crook, pp. 169–210. London & New York: Academic Press.

Archer, J. (1971). Sex differences in emotional behaviour: a reply to Gray and Buffery. *Acta Psychologica*, **35**, 415–29.

Archer, J. (1974). The effects of testosterone on the distractability of chicks by irrelevant and relevant novel stimuli. *Animal Behaviour*, **22**, 397–404.

Archer, J. (1975). Rodent sex differences in emotional and related behavior. *Behavioral Biology*, **14**, 451–79.

Archer, J. (1976a). The organization of aggression and fear in vertebrates. In *Perspectives in Ethology 2*, ed. P. P. G. Bateson & P. Klopfer, pp. 231–98. New York & London: Plenum.

Archer, J. (1976b). Testosterone and fear behavior in male chicks. *Physiology and Behavior*, **17**, 561–4.

Archer, J. (1977a). The psychology of violence. *New Society*, **42**, 63–6.

Archer, J. (1977b). Testosterone and persistence in mice. *Animal Behaviour*, **25**, 479–88.

Archer, J. (1979). Behavioural aspects of fear. In *Fear in Animals and Man*, ed. W. Sluckin, pp. 56–85. Wokingham, UK: Van Nostrand.

Archer, J. (1986a). Animal sociobiology and comparative psychology: a review. *Current Psychological Research and Reviews*, **5**, 48–61.

Archer, J. (1986b). Ethical issues in psychobiological research on animals. *Bulletin of the British Psychological Society*, **39**, 361–4.

Archer, J. (1986c). Game theoretic models and respect for ownership. *The Behavioral and Brain Sciences*, **9**, 740–4.

Archer, J. & Blackman, D. E. (1971). Prenatal psychological stress and offspring behavior in rats and mice. *Developmental Psychobiology*, **4**, 193–248.

Archer, J. & Browne, K. (in press). Concepts and approaches to the study of aggression. In *Human Aggression: Naturalistic Approaches*, ed. J. Archer & K. Browne. Beckenham, UK: Croom Helm.

Arnold, S. J. & Bennett, A. F. (1984). Behavioural variation in natural populations. III. antipredator displays in the garter snake *Thamnophis radix*. *Animal Behaviour*, **32**, 1108–18.

Arvola, A., Ilmen, M. & Koponen, T. (1962). On the aggressive behaviour of the Norwegian lemming *Lemmus lemmus* with special reference to the sounds produced. *Archivum Societatis Zoologicae Botanicae Fennicae 'Vanamo'*, **17**, 80–101.

Austad, S. N. (1983). A game theoretical interpretation of male combat in the bowl and doily spider (*Frontinella pyramitela*). *Animal Behaviour*, **31**, 59–73.

Austad, S. N. & Howard, R. D. (1984). Introduction to the symposium: alternative reproductive tactics. *American Zoologist*, **24**, 307–8.

Ayer, M. L. & Whitsett, J. M. (1980). Aggressive behaviour of female prairie deer mice in laboratory populations. *Animal Behaviour*, **28**, 763–71.

Azrin, N. H. & Hutchinson, R. R. (1967). Conditioning of the aggressive behavior of pigeons by a fixed-interval schedule of reinforcement. *Journal of the Experimental Analysis of Behavior*, **10**, 395–402.

Azrin, N. H., Hake, D. F. & Hutchinson, R. R. (1965a). Elicitation of aggression by a physical blow. *Journal of the Experimental Analysis of Behavior*, **8**, 55–7.

Azrin, N. H., Hutchinson, R. R. & McLaughlin, R. (1965b). The opportunity for aggression as an operant reinforcer during aversive stimulation. *Journal of the Experimental Analysis of Behavior*, **8**, 171–80.

Baenninger, L. P. (1973). Aggression between weanling *Peromyscus* and *Microtus*. *Animal Behaviour*, **21**, 335–7.

Baerends, G. P. (1975). An evaluation of the conflict hypothesis as an explanatory principle for the evolution of displays. In *Function and Evolution in Behaviour*, ed. G. P. Baerends, C. Beer & A. Manning, pp. 187–227. Oxford: Clarendon Press.

Baerends, G. P. (1985). Do dummy experiments with sticklebacks support the IRM-concept? *Behaviour*, **93**, 258–77.

Baggerman, B. (1966). On the endocrine control of reproductive behavior in the male three-spined stickleback *Gasterosteus aculeatus* L. *Symposia of the Society for Experimental Biology*, **20**, 427–56.

Bakker, K. (1961). An analysis of factors which determine success in competition for food among larvae of *Drosophila melanogaster*. *Archives Neerlandaises de Zoologie*, **14**, 200–81.

Bakker, T. C. M. (1985). Two-way selection for aggression in juvenile female and male sticklebacks (*Gasterosteus aculeatus* L.), with some notes on hormonal factors. *Behaviour*, **93**, 69–81.

Bakker, T. C. M. (1986). Aggressiveness in sticklebacks (*Gasterosteus aculeatus* L.): a behaviour–genetic study. *Behaviour*, **98**, 1–144.

Bandura, A. (1973a). *Aggression: A Social Learning Analysis*. New Jersey: Prentice Hall.

Bandura, A. (1973b). Social learning theory and aggression. In *The Control of Aggression*, ed. J. F. Knutson, pp. 201–50. Chicago: Aldine.

Bandura, A., Ross, D. & Ross, S. A. (1961). Transmission of aggression through imitation of aggressive models. *Journal of Abnormal and Social Psychology*, **63**, 575–82.

Barfield, R. J. (1984). Reproductive hormones and aggressive behavior. In *Biological Perspectives on Aggression*, ed. K. J. Flannelly, R. J. Blanchard & D. C. Blanchard, pp. 105–34. New York: Alan Liss.

Barkan, C. P. L., Craig, J. L., Strahl, S. D., Stewart, A. M. & Brown, J. L.

(1986). Social dominance in communal Mexican jays, *Aphelocoma ultramarina*. *Animal Behaviour*, **34**, 175–87.

Barnett, S. A. (1955). Competition among wild rats. *Nature*, **175**, 126–7.

Barnett, S. A., Evans, C. S. & Stoddart, R. C. (1958). Influence of females on conflict among wild rats. *Journal of Zoology*, **154**, 391–6.

Bateson, P. P. G. (1964). Changes in chicks' responses to novel moving objects over the sensitive period for imprinting. *Animal Behaviour*, **12**, 479–89.

Bateson, P. P. G. (1978). Sexual imprinting and optimal outbreeding. *Nature*, **273**, 659–60.

Bateson, P. P. G. (1980). Optimal outbreeding and the development of sexual preferences in Japanese Quail. *Zeitschrift für Tierpsychologie*, **53**, 231–44.

Bateson, P. P. G. (1982). Preferences for cousins in Japanese Quail. *Nature*, **295**, 236–7.

Bateson, P. P. G. (1984). Genes, environment and learning. In *The Biology of Learning*, ed. P. Marler & H. S. Terrace, pp. 75–88. Berlin & New York: Springer-Verlag.

Bateson, P. P. G. (1986). When to experiment on animals. *New Scientist*, **109** (1496), 30–2.

Batty, J. (1978*a*). Plasma levels of testosterone and male sexual behaviour in strains of the house mouse (*Mus musculus*). *Animal Behaviour*, **26**, 339–48.

Batty, J. (1978*b*). Acute changes in plasma testosterone levels and their relation to measures of sexual behaviour in the male house mouse (*Mus musculus*). *Animal Behaviour*, **26**, 349–57.

Beach, F. A. (1950). The snark was a boojum. *American Psychologist*, **5**, 115–24.

Beach, F. A. (1976). Sexual attractivity, proceptivity and receptivity in female mammals. *Hormones and Behavior*, **7**, 105–38.

Beatty, W. W. (1979). Gonadal hormones and sex differences in non-reproductive behaviors in rodents: organizational and activational influences. *Hormones and Behavior*, **12**, 112–63.

Beaugrand, J. P., Caron, J. & Comeau, L. (1984). Social organization of small heterosexual groups of green swordtails (*Xiphophorus helleri*, Pisces, Poeciliiadae) under conditions of captivity. *Behaviour*, **91**, 24–60.

Beeman, A. E. (1947*a*). The effect of male hormone on aggressive behavior in mice. *Physiological Zoology*, **20**, 373–405.

Beeman, E. A. (1947*b*). The relation of the interval between castration and first encounter to the aggressive behavior of mice. *Anatomical Record*, **99**, 570–1.

Beilharz, R. G. & Beilharz, V. C. (1975). Observations on fighting behaviour of male mice (*Mus musculus* L.). *Zeitschrift für Tierpsychologie*, **39**, 126–40.

Bekoff, M. (1981). Development of agonistic behaviour: ethological and ecological aspects. In *Multidisciplinary Approaches to Aggression Research*, ed. P. F. Brain & D. Benton, pp. 161–78. Amsterdam: Elsevier/North Holland.

Bekoff, M., Tyrrell, M., Lipetz, V. E. & Jamieson, R. (1981). Fighting patterns in young coyotes: imitation, escalation and assessment. *Aggressive Behavior*, **7**, 225–44.

Berkowitz, L. (1962). *Aggression: A Social Psychological Analysis*. New York: McGraw-Hill.

Berkowitz, L. (1969). The frustration-aggression hypothesis revisited. In *Roots of Aggression*, ed. L. Berkowitz, pp. 1–28. New York: Atherton.

Berkowitz, L. (1970). The contagion of violence: an S-R mediational analysis of some effects of observed aggression. In *Nebraska Symposium on Motivation*, pp. 95–135. Lincoln: University of Nebraska Press.

Berkowitz, L. (1983). Aversively-stimulated aggression. *American Psychologist*, **38**, 1135–44.

Berkowitz, L. (1984). Physical pain and the inclination to aggression. In *Biological Perspectives on Aggression*, ed. K. J. Flannelly, R. J. Blanchard & D. C. Blanchard, pp. 27–47. New York: Liss.

Bernstein, I. S. (1981). Dominance: the baby and the bathwater. *The Behavioral and Brain Sciences*, **4**, 419–29.

Bernstein, I. S. & Gordon, T. P. (1974). The function of aggression in primate societies. *American Scientist*, **62**, 304–11.

Bernstein, I., Gordon, T. P. & Rose, R. M. (1983). The interaction of hormones, behavior and social context in nonhuman primates. In *Hormones and Aggressive Behavior*, ed. B. B. Svare, pp. 535–61. New York: Plenum.

Berrill, M. & Arsenault, M. (1982). Mating behavior of the green shore crab *Carcinus maenas*. *Bulletin of Marine Science*, **32**, 632–8.

Berrill, M. & Arsenault, M. (1984). The breeding behaviour of a northern temperate Orconectid crayfish, *Orconectes rusticus*. *Animal Behaviour*, **32**, 333–9.

Bertram, B. C. R. (1975). Social factors influencing reproduction in wild lions. *Journal of Zoology*, **177**, 463–82.

Bevan, W., Daves, W. F. & Levy, G. W. (1960). The relation of castration, androgen therapy and pre-test fighting experience to competitive aggression in male *C57Bl/10* mice. *Animal Behaviour*, **8**, 6–12.

Bishop, D. T. & Cannings, C. (1978). A generalized War of Attrition. *Journal of Theoretical Biology*, **70**, 85–124.

Bishop, D. T., Cannings, C. & Maynard Smith, J. (1978). The War of Attrition with random rewards. *Journal of Theoretical Biology*, **74**, 377–88.

Black-Cleworth, P. (1970). The role of electrical discharges in the non-reproductve social behaviour of *Gymnotus carapo* (Gymnotidae, Pisces). *Animal Behaviour Monographs*, **3**, 1–77.

Blanchard, D. C., Blanchard, R. J., Takahashi, T. & Suzuki, N. (1978). Aggressive behaviors of the Japanese brown bear. *Aggressive Behavior*, **4**, 31–41.

Blanchard, D. C. & Blanchard, R. J. (1984*a*). Inadequacy of pain-aggression hypothesis revealed in naturalistic settings. *Aggressive Behavior*, **10**, 33–46.

Blanchard, D. C. & Blanchard, R. J. (1984*b*). Affect and aggression: an animal model applied to human behavior. In *Advances in the Study of Aggression*, vol. 1, ed. R. J. Blanchard & D. C. Blanchard, pp. 1–62. New York: Academic Press.

Blanchard, D. C., Fukunaga-Stinson, C., Takahashi, L. K., Flannelly, K. J. & Blanchard, R. J. (1984). Dominance and aggression in social groups of male and female rats. *Behavioural Processes*, **9**, 31–48.

Blanchard, R. J. (1984). Pain and aggression reconsidered. In *Biological Perspectives on Aggression*, ed. K. J. Flannelly, R. J. Blanchard & D. C. Blanchard, pp. 1–26. New York: Liss.

Blanchard, R. J. & Blanchard, D. C. (1977). Aggressive behavior in the rat. *Behavioral Biology*, **21**, 197–224.

Blanchard, R. J. & Blanchard, D. C. (1981). The organization and modeling of animal aggression. In *The Biology of Aggression*, ed. P. F. Brain & D. Benton, pp. 529–61. Rockville, Maryland: Sijthoff & Noordhoff.

Blanchard, R. J., Flannelly, K. J., Layng, M. & Blanchard, D. C. (1984*a*). The effects of age and strain on aggression in male rats. *Physiology and Behavior*, **33**, 857–61.

Blanchard, R. J., Kleinschmidt, C. K., Flannelly, K. J. & Blanchard, D. C. (1984*b*). Fear and aggression in the rat. *Aggressive Behavior*, **10**, 309–16.

Blanchard, R. J., Mast, M. & Blanchard, D. C. (1975). Stimulus control of defensive reactions in the albino rat. *Journal of Comparative and Physiological Psychology*, **88**, 81–8.

Blancher, P. J. & Robertson, R. J. (1982). Kingbird aggression: does it deter predation? *Animal Behaviour*, **30**, 929–30.

Blest, D. (1961). The concept of ritualization. In *Current Problems in Animal Behaviour*, ed. W. H. Thorpe & O. L. Zangwill, pp. 102–24. London & New York: Cambridge University Press.

Blurton-Jones, N. G. (1968). Observations and experiments on causation of threat displays of the Great Tit *Parus major*. *Animal Behaviour Monographs*, **1**, 75–158.

Bolles, R. C. (1970). Species-specific defence reactions and avoidance learning. *Psychological Review*, **77**, 32–48.

Bols, R. J. (1977). Display reinforcement in the Siamese fighting fish, *Betta splendens*: aggressive motivation or curiosity? *Journal of Comparative and Physiological Psychology*, **91**, 233–44.

Borradaile, L. A., Eastham, L. E. S., Potts, F. A. & Saunders, J. T. (1961). *The Invertebrata* (4th edn), revised by G. A. Kerkut. Cambridge: Cambridge University Press.

Boska, S. C., Weisman, H. M. & Thor, D. H. (1966). A technique for inducing aggression in rats utilizing morphine withdrawal. *Psychological Record*, **16**, 541–3.

Bouissou, M-F. (1983). Hormonal influences on aggressive behavior in ungulates. In *Hormones and Aggressive Behavior*, ed. B. B. Svare, pp. 507–33. New York: Plenum.

Bowlby, J. (1969). *Attachment*. London: Hogarth Press & Penguin.

Bowlby, J. (1973). *Separation: Anxiety and Anger*. London: Hogarth Press.

Brace, R. C. & Pavey, J. (1978). Size-dependent dominance hierarchy in the anemone *Actinia equina*. *Nature*, **273**, 752–3.

Brace, R. C., Pavey, J. & Quicke, D. L. J. (1979). Intraspecific aggression in the colour morphs of the anemone *Actinia equina*: the 'convention' governing dominance ranking. *Animal Behaviour*, **27**, 553–61.

Brain, P. F. (1977). *Hormones and Aggression*. Annual Research Reviews Vol. 1. Montreal: Eden Press; Edinburgh: Churchill-Livingstone.

Brain, P. F. (1979). *Hormones and Aggression*. Annual Research Reviews Vol. 2. Montreal: Eden Press; Edinburgh: Churchill-Livingstone.

Brain, P. F. (1981*a*). Hormones and aggression in infra-human vertebrates. In *The Biology of Aggression*, ed. P. F. Brain & D. Benton, pp. 181–213. Rockville, Maryland: Sijthoff & Noordhoff.

Brain, P. F. (1981*b*). Differentiating types of attack and defense in rodents. In *Multidisciplinary Approaches to Aggression Research*, ed. P. F. Brain & D. Benton, pp. 53–78. Amsterdam: Elsevier/North Holland.

Brain, P. F., Al-Malki, S. & Benton, D. (1981). Attempts to determine the status of electroshock-induced attack in male laboratory mice. *Behavioural Processes*, **6**, 171–89.

Brain, P. F. & Benton, D. (1983). Conditions of housing, hormones, and aggressive behavior. In *Hormones and Aggressive Behavior*, ed. B. B. Svare, pp. 351–72. New York: Plenum.

Brain, P. F., Nowell, N. W. & Wouters, A. (1971). Some relationships between adrenal function and the effectiveness of a period of isolation in inducing intermale aggression in albino mice. *Physiology and Behavior*, **6**, 27–9.

Brawn, V. M. (1960). Aggressive behaviour in the cod *Gadus callarias* L. *Behaviour*, **18**, 107–47.

Breed, M. D. & Bell, W. J. (1983). Hormonal influences on invertebrate aggressive behavior. In *Hormones and Aggressive Behavior*, ed. B. B. Svare, pp. 577–90. New York: Plenum.

Breed, M. D., Hinkle, C. M. & Bell, W. J. (1975). Agonistic behavior in the German cockroach (*Blattella germanica*). *Zeitschrift für Tierpsychologie*, **39**, 24–32.

Bronstein, P. M. (1981). Commitments to aggression and nest sites in male *Betta splendens*. *Journal of Comparative and Physiological Psychology*, **95**, 436–9.

Bronstein, P. M. (1983). Agonistic sequences and the assessment of opponents in male *Betta splendens*. *American Journal of Psychology*, **96**, 163–77.

Bronstein, P. M. (1984). Agonistic and reproductive interactions in *Betta splendens*. *Journal of Comparative Psychology*, **98**, 421–31.

Bronstein, P. M. (1985*a*). Prior-residence effect in *Betta splendens*. *Journal of Comparative Psychology*, **99**, 56–9.

Bronstein, P. M. (1985*b*). Predictors of dominance in male *Betta splendens*. *Journal of Comparative Psychology*, **99**, 47–55.

Brown, J. A. & Colgan, P. W. (1985). The ontogeny of social behaviour in four species of centrarchid fish. *Behaviour*, **92**, 254–76.

Brown, J. H. & Lasiewski, R. C. (1972). Metabolism of weasels: the cost of being long and thin. *Ecology*, **53**, 939–43.

Brown, J. L. (1964). The evolution of diversity in avian territorial systems. *Wilson Bulletin*, **6**, 160–9.

Brown, R. & Herrnstein, R. J. (1975). *Psychology*. Boston: Little Brown.

Bruce, H. M. & Parrott, D. M. V. (1960). Role of olfactory sense in pregnancy block by strange males. *Science*, **131**, 1526.

Bujalska, G. (1973). The role of spacing behaviour among females in the regulation of reproduction in the bank vole. *Journal of Reproduction and Fertility*, Supplement **19**, 465–74.

Burghardt, G. M. (1978). Behavioral ontogeny in reptiles: whence, wither and why? In *The Development of Behavior: Comparative and Evolutionary Aspects*, ed. G. M. Burghardt & M. Bekoff, pp. 149–74. New York & London: Garland STPM Press.

Burton, M. (1969). *The Hedgehog*. London: Deutsch.

Buss, A. H. (1971). Aggression pays. In *The Control of Aggression and Violence*, ed. J. L. Singer, pp. 7–18. New York & London: Academic Press.

Butler, R. G. (1980). Population size, social behaviour, and dispersal in house mice: a quantitative investigation. *Animal Behaviour*, **28**, 78–85.

Bygott, J. D. (1972). Cannibalism among wild chimpanzees. *Nature*, **238**, 410–11.

Cairns, R. B., Hood, K. E. & Midlam, J. (1985). On fighting in mice: is there a sensitive period for isolation effects? *Animal Behaviour*, **33**, 166–80.

Calhoun, J. B. (1963). *The Ecology and Sociology of the Norway Rat*. Bethesda, Maryland: U.S. Department of Health, Education, & Welfare, Public Health Service.

Campagnoni, F. R., Cohen, P. S., & Yoburn, B. C. (1981). Organization of attack and other behaviors of White King pigeons exposed to intermittent water presentations. *Animal Learning and Behavior*, **9**, 491–500.

Carlisle, T. R. (1982). Brood success in variable environments: implications for parental care allocation. *Animal Behaviour*, **30**, 824–36.

Carlisle, T. R. (1985). Parental response to brood size in a cichlid fish. *Animal Behaviour*, **33**, 234–8.

Carpenter, F. L. & Macmillan, R. E. (1976). Threshold model of feeding territoriality and test with a Hawaiian honeycreeper. *Science*, **194**, 639–42.

Carpenter, F. L., Paton, D. C., & Hixon, M. A. (1983). Weight gain and adjustment of feeding territory size in migrant hummingbirds. *Proceedings of the National Academy of Sciences*, **80**, 7259–63.

Caryl, P. G. (1979). Communication by agonistic displays: what can game theory contribute to ethology? *Behaviour*, **68**, 136–69.

Caryl, P. G. (1981). Escalated fighting and the war of nerves: games theory and animal combat. In *Perspectives in Ethology* 4, ed. P. P. G. Bateson & P. Klopfer, pp. 199–224. New York & London: Plenum.

Chalmers, N. R. (1968). The social behaviour of free living mangabeys in Uganda. *Folia Primatologica*, **8**, 263–81.

Chalmers, N. R. (1981). Dominance as part of a relationship. *The Behavioral and Brain Sciences*, **4**, 437–8.

Chantry, D. & Workman, L. (1984). Song and plumage effects on aggressive display by the European robin *Erithacus rubecula*. *The Ibis*, **126**, 366–71.

Chelazzi, G., Focardi, S., Deneubourg, J. L. & Innocenti, R. (1983). Competition for the home and aggressive behavior in the chiton *Acanthopleura gemmata* (Blainville) (Mollusca: Polyplacophora). *Behavioral Ecology and Sociobiology*, **14**, 15–20.

Chitty, D. (1952). Mortality among voles (*Microtus agrestis*) at Lake Vyrnwy, Montgomeryshire in 1936–9. *Philosophical Transactions of the Royal Society B*, **236**, 505–52.

Chitty, D. (1960). Population processes in the vole and their relevance to general theory. *Canadian Journal of Zoology*, **38**, 99–113.

Christenson, T. E. & Goist, K. C. Jr. (1979). Costs and benefits of male–male competition in the orb weaving spider, *Nephila clavipes*. *Behavioral Ecology and Sociobiology*, **5**, 87–92.

Clarke, J. R. (1956). The aggressive behaviour of the vole. *Behaviour*, **9**, 1–23.

Clausen, C. P. (1940). *Entomophagous Insects*. New York: McGraw-Hill.

Clifton, P. G., Andrew, R. J. & Rainey, C. R. (1986). Effects of gonadal steroids on attack and on memory processing in the domestic chick. *Physiology and Behavior*, **37**, 701–7.

Cloudsley-Thompson, J. L. (1980). *Tooth and Claw*. London: Dent.

Clutton-Brock, T. H. (1974). Primate social organisation and ecology. *Nature*, **250**, 539–42.

Clutton-Brock, T. H. & Albon, S. D. (1979). The roaring of red deer and the evolution of honest advertisement. *Behaviour*, **69**, 145–70.

Clutton-Brock, T. H., Albon, S. D., Gibson, R. M. & Guinness, F. E. (1979). The logical stag: adaptive aspects of fighting in red deer (*Cervas elaphas* L.). *Animal Behaviour*, **27**, 211–25.

Clutton-Brock, T. H., Guinness, F. E. & Albon, S. D. (1982). *Red Deer: Behavior and Ecology of Two Sexes*. Edinburgh: Edinburgh University Press.

Clutton-Brock, T. H. & Harvey, P. H. (1976). Evolutionary rules and primate societies. In *Growing Points in Ethology*, ed. P. P. G. Bateson & R. A. Hinde, pp. 195–237. Cambridge: Cambridge University Press.

Coleman, R. M., Gross, M. R. & Sargent, R. C. (1986). Parental investment decision rules: a test in bluegill sunfish. *Behavioral Ecology and Sociobiology*, **18**, 59–66.

Colgan, P. W., Nowell, W. A. & Stokes, N. W. (1981). Spatial aspects of nest defence by pumpkinseed sunfish (*Lepomis gibbosus*): stimulus features and an application of catastrophe theory. *Animal Behaviour*, **29**, 433–42.

Collier, G., Hirsch, E. & Hamlin, P. H. (1973). The ecological determinants of reinforcement in the rat. *Physiology and Behavior*, **9**, 705–16.

Conder, P. J. (1949). Individual distance. *The Ibis*, **91**, 649–55.

Connell, J. H. (1961). The influence of interspecific competition and other factors on the distribution of the barnacle *Chthalamus stellatus*. *Ecology*, **42**, 710–23.

Connell, J. H. (1963). Territorial behavior and dispersion in some marine invertebrates. *Researches on Population Ecology*, **5**, 87–101.

Conner, R. L., Constantino, A. P. & Scheuch, G. C. (1983). Hormonal influences on shock-induced fighting. In *Hormones and Aggressive Behavior*, ed. B. B. Svare, pp. 119–44. New York: Plenum.

Craig, J. L. & Douglas, M. E. (1984). Temporal partitioning of a nectar resource in relation to competitive asymmetries. *Animal Behaviour*, **32**, 624–9.

Craig, J. L. & Douglas, M. E. (1986). Resource distribution, aggressive asymmetries and variable access to resources in the nectar feeding bellbird. *Behavioral Ecology and Sociobiology*, **18**, 231–40.

Craig, W. (1918). Appetites and aversions as constituents of instincts. *Biological Bulletin*, **34**, 91–107.

Craig, W. (1928). Why do animals fight? *International Journal of Ethics*, **31**, 264–78.

Creer, T. L., Hitzing, E. W. & Schaeffer, R. W. (1966). Classical conditioning of reflexive fighting. *Psychonomic Science*, **4**, 89–90.

Crespi, B. J. (1986). Size assessment and alternative fighting tactics in *Elaphrothrips tuberculatus* (Insecta: Thysanoptera). *Animal Behaviour*, **34**, 1324–35.

Croll, J. (1976). Is catastrophe theory dangerous? *New Scientist*, **70**, 630–2.

Crook, J. H. (1965). The adaptive significance of avian social organisations. *Symposium of the Zoological Society of London*, **14**, 181–218.

218   *References*

Crook, J. H. (1970a). Social organization and the environment: aspects of contemporary social ethology. *Animal Behaviour*, **18**, 197–209.

Crook, J. H. (1970b). The socio-ecology of primates. In *Social Behaviour in Birds and Mammals*, ed. J. H. Crook, pp. 103–66. London & New York: Academic Press.

Crook, J. H. & Butterfield, P. A. (1968). Effects of testosterone propionate and luteinizing hormone on agonistic and nest building behaviour of *Quelea quelea*. *Animal Behaviour*, **16**, 370–84.

Crook, J. H. & Gartlan, J. S. (1966). Evolution of primate societies. *Nature*, **210**, 1200–3.

Crosby, R. M. & Cahoon, D. D. (1973). Superstitious responding as an artifact in investigations of shock-elicited aggression. *Psychological Record*, **23**, 191–6.

Crowcroft, P. (1955). Territoriality in wild house mice, *M. musculus*. *Journal of Mammalogy*, **36**, 299–301.

Curio, E. (1973). Towards a methodology of teleonomy. *Experientia*, **29**, 1045–58.

Curio, E. (1975). The functional organization of anti-predator behaviour in the Red Flycatcher: a study of avian visual perception. *Animal Behaviour*, **23**, 1–115.

Curio, E., Regelmann, K. & Zimmermann, U. (1984). The defence of first and second broods by great tit (*Parus major*) parents: a test of predictive sociobiology. *Zeitschrift für Tierpsychologie*, **66**, 101–27.

Curio, E., Regelmann, K. & Zimmermann, U. (1985). Brood defence in the great tit (*Parus major*): the influence of life-history and habitat. *Behavioral Ecology and Sociobiology*, **16**, 273–83.

Curry-Lindahl, K. (1962). The irruption of the Norway lemming in Sweden during 1960. *Journal of Mammalogy*, **43**, 171–84.

Darwin, C. (1871). *The Descent of Man, and Selection in Relation to Sex* (1901 edition). London: Murray.

Davies, N. B. (1978a). Ecological questions about territorial behaviour. In *Behavioural Ecology: An Evolutionary Approach*, ed. J. R. Krebs & N. B. Davies, pp. 371–50. Oxford: Blackwell Scientific.

Davies, N. B. (1978b). Territorial defence in the speckled wood butterfly (*Pararge aegeria*): the resident always wins. *Animal Behaviour*, **26**, 138–47.

Davies, N. B. (1981). Calling as an ownership convention on pied wagtail territories. *Animal Behaviour*, **29**, 529–34.

Davies, N. B. (1985). Cooperation and conflict among dunnocks, *Prunella modularis*, in a variable mating system. *Animal Behaviour*, **33**, 628–48.

Davies, N. B. & Halliday, T. M. (1978). Deep croaks and fighting assessment in toads *Bufo bufo*. *Nature*, **274**, 683–5.

Davies, N. B. & Houston, A. I. (1984). Territory economics. In *Behavioural Ecology: An Evolutionary Approach* (2nd edn), ed. J. R. Krebs & N. B. Davies, pp. 148–69. Oxford: Blackwell Scientific.

Davies, N. B. & Lundberg, A. (1984). Food distribution and a variable mating system in the dunnock, *Prunella modularis*. *Journal of Animal Ecology*, **53**, 895–913.

Davis, D. E. (1957). Aggressive behavior in castrated starlings. *Science*, **126**, 253.

Davis, J. M. (1973). Imitation: a review and critique. In *Perspectives in Ethology*, P. P. G. Bateson & P. Klopfer, pp. 43–72. New York: Plenum.

Davis, R. E. & Kassel, J. K. (1975). The ontogeny of agonistic behavior and the onset of sexual maturation in the Paradise fish, *Macropodus opercularis* (Linnaeus). *Behavioral Biology*, **14**, 31–9.

Davis, W. M. & Khalsa, J. H. (1971). Some determinants of aggressive behavior induced by morphine withdrawal. *Psychonomic Science*, **24**, 13–15.

Dawkins, R. (1976). *The Selfish Gene*. Oxford: Oxford University Press.

Dawkins, R. & Krebs, J. R. (1978). Animal signals: information or manipulation. In *Behavioural Ecology*, ed. J. R. Krebs & N. B. Davies, pp. 282–309. Oxford: Blackwell Scientific.

Dawkins, R. & Carlisle, T. R. (1976). Parental investment, mate desertion and a fallacy. *Nature*, **262**, 131–3.

Day, H. D., Seay, B. M., Hale, P. & Hendricks, D. (1982). Early social deprivaton and the ontogeny of unrestricted social behavior in the laboratory rat. *Developmental Psychobiology*, **15**, 47–59.

De Boer, J. N. & Heuts, B. (1973). Prior exposure to visual cues affecting dominance in the jewel fish, *Hemichromis bimaculatus* Gill 1862 (Pisces, Cichlidae). *Behaviour*, **44**, 299–321.

DeBold, J. F. & Miczek, K. A. (1984). Aggression persists after ovariectomy in female rats. *Hormones and Behavior*, **18**, 177–90.

De Ghett, V. J. (1975). A factor influencing aggression in adult mice: witnessing aggression when young. *Behavioral Biology*, **13**, 291–300.

De Jonge, G. (1983). Aggression and group formation in voles *Microtus agrestis*, *M. arvalis*, and *Clethrionomys glareolus*, in relation to intra- and interspecific competition. *Behaviour*, **84**, 1–73.

De Waal, F. B. M. (1984). Sex differences in the formation of coalitions among chimpanzees. *Ethology and Sociobiology*, **5**, 239–55.

Denenberg, V. H. (1973). Developmental factors in aggression. In *The Control of Aggression*, ed. J. F. Knutson, pp. 41–57. Chicago: Aldine.

Dewsbury, D. A. (1984*a*). *Comparative Psychology in the Twentieth Century*. Stroudsburg, Pennsylvania: Hutchinson Ross.

Dewsbury, D. A. (1984*b*). Aggression, copulation and differential reproduction of deer mice (*Peromyscus maniculatus*) in a semi-natural enclosure. *Behaviour*, **91**, 1–23.

Dixon, K. A. & Cade, W. H. (1986). Some factors influencing male–male aggression in the field cricket *Gryllus integer* (time of day, age, weight and sexual maturity). *Animal Behaviour*, **34**, 340–6.

Dollard, J., Doob, L. W., Miller, N. E., Mowrer, O. H. & Sears, R. R. (1939). *Frustration and Aggression*. New Haven: Yale University Press.

Dow, M., Ewing, A. W. & Sutherland, I. (1976). Studies on the behaviour of Cyprinodont fish. III. The temporal patterning of aggression in *Aphyosemion striatum* (Boulenger). *Behaviour*, **59**, 252–68.

Dowds, B. M. & Elwood, R. W. (1983). Shell wars: assessment strategies and the timing of decisions in hermit crab shell fights. *Behaviour*, **85**, 1–24.

Dowds, B. M. & Elwood, R. W. (1985). Shell wars II: the influence of relative size on decisions made during hermit crab shell fights. *Animal Behaviour*, **33**, 649–56.

Drewett, R. & Kani, W. (1981). Animal experimentation in the behavioural sciences. In *Animals in Research*, ed. D. Sperlinger. Chichester: Wiley.

Dudzinski, M. L., Mykytowycz, R. & Gambale, S. (1977). Behavioral characteristics of adolescence in young captive European rabbits, *Oryctolagus cuniculus*. *Aggressive Behavior*, 3, 313–30.

Duncan, P., Harvey, P. H. & Wells, S. M. (1984). On lactation and associated behaviour in a natural herd of horses. *Animal Behaviour*, 32, 255–63.

Duval, S. & Wicklund, R. A. (1972). *A Theory of Objective Self Awareness*. New York: Academic Press.

Ebert, P. D. (1983). Selection for aggression in a natural population. In *Aggressive Behavior: Genetic and Neural Approaches*, ed. E. C. Himmel, M. E. Hahn & J. K. Walters, pp. 103–27. Hillsdale, New Jersey: Lawrence Erlbaum.

Ebert, P. D. & Hyde, J. S. (1976). Selection for agonistic behavior in wild female *Mus musculus*. *Behavior Genetics*, 6, 291–304.

Edmunds, G. & Kendrick, D. C. (1980). *The Measurement of Human Aggressiveness*. Chichester: Ellis Harwood.

Edmunds, M. (1972). Defensive behaviour in Ghanaian praying mantids. *Zoological Journal of the Linnean Society*, 51, 1–32.

Edmunds, M. (1974). *Defence in Animals: A Survey of Anti-Predator Defences*. Harlow: Longman.

Edmunds, M. (1976). The defensive behaviour of Ghanaian praying mantids with a discussion of territoriality. *Zoological Journal of the Linnean Society*, 58, 1–37.

Edwards, D. A. (1969). Early androgen stimulation and aggressive behavior in male and female mice. *Physiology and Behavior*, 40, 333–8.

Eibl-Eibesfeldt, I. (1961). The fighting behavior of animals. *Scientific American*, 205, 112–22.

Einon, D. F. (1980). The purpose of play. In *Not Work Alone*, ed. J. Cherfas & R. Lewin, pp. 21–32. London: Temple Smith.

Einon, D. F. (1983). Play and exploration. In *Exploration in Animals and Humans*, ed. J. Archer & L. I. A. Birke, pp. 210–29. Wokingham U.K.: Van Nostrand.

Eisenberg, J. F. (1962). Studies on the behavior of *Peromyscus maniculatus gambelii* and *Peromyscus californicus parastiticus*. *Behaviour*, 19, 177–207.

Eisenberg, J. F. (1981). Parental care. In *The Oxford Companion to Animal Behaviour*, ed. D. McFarland, pp. 443–7. Oxford: Oxford University Press.

Elton, C. (1942). *Voles, Mice and Lemmings*. London: Oxford University Press.

Elwood, R. W. (1980). The development, inhibition and disinhibition of pup-cannibalism in the mongolian gerbil. *Animal Behaviour*, 28, 1188–94.

Elwood, R. W. (1985). Inhibition of infanticide and onset of paternal care in male mice (*Mus musculus*). *Journal of Comparative Psychology*, 99, 457–67.

Emlen, S. T. & Oring, L. W. (1977). Ecology, sexual selection and the evolution of mating systems. *Science*, 197, 215–23.

Enquist, M. (1985). Communication during aggressive interactions with particular reference to variation in choice of behaviour. *Animal Behaviour*, 33, 1152–61.

Enquist, M. & Leimar, O. (1983). Evolution of fighting behaviour: decision rules and assessment of relative strength. *Journal of Theoretical Biology*, **102**, 387–410.

Epple, G. (1981). Effect of pair-bonding with adults on the ontogenetic manifestation of aggressive behavior in a primate *Saguinus fuscicollis*. *Behavioral Ecology and Sociobiology*, **8**, 117–23.

Erskine, M. S., Denenberg, V. H., Goldman, B. D. (1978). Aggression in the lactating rat: effects of intruder age and test arena. *Behavioral Biology*, **23**, 52–66.

Evans, R. M. (1967). Early aggressive responses in domestic chicks. *Animal Behaviour*, **16**, 24–8.

Evans, S. (1983). The pair-bond of the common marmoset, *Callithrix jacchus jacchus*: an experimental investigation. *Animal Behaviour*, **31**, 651–8.

Evans, S. M. (1973). A study of fighting reactions in some Nereid polychaetes. *Animal Behaviour*, **21**, 138–46.

Ewald, E. W. (1985). Influence of asymmetries in resource quality and age on aggression and dominance in black-chinned hummingbirds. *Animal Behaviour*, **33**, 705–19.

Ewer, R. F. (1969). *Ethology of Mammals*. London: Elek Science.

Fairbanks, L. A. & McGuire, M. T. (1986). Age, reproductive value, and dominance-related behaviour in vervet monkey females: cross-generational influences on social relationships and reproduction. *Animal Behaviour*, **34**, 1710–21.

Fantino, E. & Logan, C. A. (1976). *The Experimental Analysis of Behavior: A Biological Perspective*. San Francisco: Freeman.

Ferguson, G. W. (1966). Releasers of courtship and territorial behaviour in the side-blotched lizard, *Uta stansburiana*. *Animal Behaviour*, **14**, 89–92.

Feshbach, S. (1964). The function of aggression and the regulation of aggressive drive. *Psychological Review*, **71**, 257–72.

Feshbach, S., Stiles, W. B. & Ditter, E. (1967). The reinforcing effect of witnessing aggression. *Journal of Experimental Research in Personality*, **2**, 133–9.

Figler, M. H., Dyer, R. S., Streckfus, C. F. & Nardini, K. M. (1975). The establishment of dominance relationships in the jewel fish *Hemichromis bimaculatus* (Gill), as related to prior exposure to and configuration of visual cues. *Behavioral Biology*, **14**, 241–5.

Figler, M. H. & Einhorn, D. M. (1983). The territorial prior residence effect in convict cichlids *Cichlasoma nigrofasciatum* (Gunther): temporal aspects of establishment and retention, and proximate mechanisms. *Behaviour*, **85**, 157–83.

Figler, M. H., Klein, R. M. & Peeke, H. V. S. (1976). The establishment and reversibility of dominance relationships in jewel fish, *Hemichromis bimaculatus* Gill (Pisces, Cichlidae): effects of prior exposure and prior residence situations. *Behaviour*, **58**, 254–71.

Fink, L. S. (1986). Costs and benefits of maternal behaviour in the green lynx spider (Oxyopidae, *Pencetia viridans*). *Animal Behaviour*, **34**, 1051–60.

Fisher, R. A. (1930). *The Genetical Theory of Natural Selection*. Oxford: Oxford University Press.

Flannelly, K. J. & Blanchard, R. J. (1981). Dominance: cause or description of social relationships. *The Behavioral and Brain Sciences*, **4**, 438–40.

Flannelly, K. J., Blanchard, R. J., Muraoke, M. Y., & Flannelly, L. (1982). Copulation increases offensive attack in male rats. *Physiology and Behavior*, **29**, 381–5.

Flannelly, K. J. & Lore, R. (1977). The influence of females upon aggression in domesticated male rats (*Rattus norvegicus*). *Animal Behaviour*, **25**, 654–9.

Flannelly, K. J. & Thor, D. H. (1978). Territorial aggression of the rat to males castrated at various ages. *Physiology and Behavior*, **20**, 785–9.

Floody, O. R. (1983). Hormones and aggression in female mammals. In *Hormones and Aggressive Behavior*, ed. B. B. Svare, pp. 39–90. New York: Plenum.

Follick, M. J. & Knutson, J. F. (1978). Punishment of irritable aggression. *Aggressive Behavior*, **4**, 1–17.

Forester, D. C. (1979). The adaptiveness of parental care in *Desmognathus ochrophaeus* (Urodela: Plethodontidae). *Copeia*, **2**, 232–41.

Francis, L. (1973). Intraspecific aggression and its effect on the distribution of *Anthopleura elegantissima* and some related sea anemones. *Biological Bulletin*, **144**, 73–92.

Frank, F. (1957). The causality of microtine cycles in Germany. *Journal of Wildlife Management*, **21**, 113–21.

Fredericson, E. & Birnbaum, E. A. (1954). Competitive fighting between mice with different hereditary backgrounds. *Journal of Genetic Psychology*, **85**, 271–80.

Freeman, D. C., Klikoff, L. G., & Harper, K. T. (1976). Differential resource utilization by the sexes of dioecious plants. *Science*, **193**, 597–9.

Fricke, H. W. (1980). Control of different mating systems in a coral reef fish by one environmental factor. *Animal Behaviour*, **28**, 561–9.

Frost, S. K. & Frost, P. G. H. (1980). Territoriality and changes in resource use by sunbirds at *Leonotis leonurus* (Labiatae). *Oecologia* (Berlin), **45**, 109–16.

Fuller, J. L. (1967). Experiential deprivation and later behavior. *Science*, **158**, 1645–52.

Galef, B. G. (1970). Aggression and timidity: responses to novelty in feral Norway rats. *Journal of Comparative and Physiological Psychology*, **70**, 370–81.

Gartlan, J. S. (1968). Structure and function in primate society. *Folia Primatologica*, **8**, 89–120.

Geist, V. (1966). The evolution of horn-like organs. *Behaviour*, **27**, 175–214.

Geist, V. (1974a). On the relationship of social evolution and ecology in ungulates. *American Zoologist*, **14**, 205–20.

Geist, V. (1974b). On fighting strategies in animal conflict. *Nature*, **250**, 354.

Geist, V. (1978). On weapons, combat and ecology. In *Advances in the Study of Communication and Affect*, vol. 4., *Aggression, Dominance and Individual Spacing*, ed. L. Krames, P. Pliner & T. Alloway, pp. 1–30. New York: Plenum.

Gentry, W. D. & Schaeffer, R. W. (1969). The effect of FR response requirement on aggressive behavior in rats. *Psychonomic Science*, **14**, 236–8.

Giles, N. (1984). Implications of parental care of offspring for the antipredator behaviour of adult male and female three-spined sticklebacks, *Gasterosteus aculeatus* L. In *Fish Reproduction: Strategies and Tactics*, ed. G. W. Potts, pp. 257–89. London: Academic Press.

Giles, N. & Huntingford, F. A. (1984). Predation-risk and inter-population variation in anti-predator behaviour in the three-spined stickleback, *Gasterosteus aculeatus* L. *Animal Behaviour*, **32**, 264–75.

Gill, F. B. & Wolf, L. L. (1975). Economics of feeding territoriality in the goldenwinged sunbird. *Ecology*, **56**, 333–45.

Ginsburg, B. & Allee, W. C. (1942). Some effects of conditioning on social dominance and subordination in inbred strains of mice. *Physiological Zoology*, **15**, 485–506.

Giordano, A. L., Siegel, H. I. & Rosenblatt, J. S. (1984). Effects of mother–litter separation and reunion on maternal aggression and pup mortality. *Physiology and Behavior*, **33**, 903–6.

Giordano, A. L., Siegel, H. I. & Rosenblatt, J. S. (1986). Intrasexual aggression during pregnancy and the estrous cycle in golden hamsters (*Mesocricetus auratus*). *Aggressive Behavior*, **12**, 213–22.

Gleason, P. E., Michael, S. D. & Christian, J. J. (1979). Effects of gonadal steroids on agonistic behavior of female *Peromyscus leucopus*. *Hormones and Behavior*, **12**, 30–9.

Gleason, P. E., Michael, S. D. & Christian, J. J. (1980). Aggressive behavior during the reproductive cycle in female *Peromyscus leucopus*: effect of encounter size. *Behavioral and Neural Biology*, **29**, 506–11.

Godfrey, G. K. (1955). Observations on the nature of the decline in numbers of two *Microtus* populations. *Journal of Mammalogy*, **36**, 209–14.

Goldman, L. & Swanson, H. H. (1975). Developmental changes in behavior (particularly aggression) from birth to maturity in confined colonies of golden hamsters. *Developmental Psychobiology*, **8**, 137–50.

Goldsmith, J. F., Brain, P. F. & Benton, D. (1976). Effects of age of differential housing and duration of individual housing/grouping on intermale fighting behavior and adrenocortical activity in T. O. strain mice. *Aggressive Behavior*, **2**, 307–23.

Gottier, R. F. (1972). Factors affecting agonistic behavior in several subhuman species. *Genetic Psychology Monographs*, **86**, 177–218.

Gould, S. J. & Lewontin, R. C. (1979). The spandrels of San Marco and the Panglossian paradigm: a critique of the adaptationist programme. *Proceedings of the Royal Society of London B*, **205**, 581–98.

Gould, S. J. & Vrba, E. S. (1982). Exaptation – a missing term in the science of form. *Paleobiology*, **8**, 4–15.

Gowaty, P. A. (1982). Sexual terms in sociobiology: emotionally evocative and, paradoxically, jargon. *Animal Behaviour*, **30**, 630–1.

Goyens, J. & Noirot, E. (1977). Intruders of differing reproductive status alter aggression differentially in early and late pregnant mice. *Aggressive Behavior*, **3**, 119–25.

Grant, E. C. & Mackintosh, J. H. (1963). A comparison of the social postures of some common laboratory rodents. *Behaviour*, **21**, 246–59.

Gray, J. A. (1971). Sex differences in emotional behaviour in mammals including man: endocrine bases. *Acta Psychologica*, **35**, 29–46.

Gray, J. A. (1979). Emotionality in male and female rodents: a reply to Archer. *British Journal of Psychology*, **70**, 425–40.

Gray, J. A. & Buffery, A. W. H. (1971). Sex differences in emotional and cognitive behaviour in mammals including Man: adaptive and neural bases. *Acta Psychologica*, **35**, 89–111.

Gray, L. E. Jr., Whitsett, J. M. & Ziesenis, J. S. (1978). Hormonal regulation of aggression towards juveniles in female house mice. *Hormones and Behavior*, **11**, 310–22.

Greenberg, N. (1983). Central and autonomic aspects of aggression in reptiles. In *Advances in Vertebrate Neuroethology*, ed. J. P. Ewert, R. R. Capranica & D. J. Ingle, pp. 1135–43. New York: Plenum.

Greenberg, N. & Crews, D. (1983). Physiological ethology of aggression in amphibians and reptiles. In *Hormones and Aggressive Behavior*, ed. B. B. Svare, pp. 469–506. New York: Plenum.

Gross, M. R. (1985). Disruptive selection for alternative life histories in salmon. *Nature*, **313**, 47–8.

Gross, M. R. & Shine, R. (1981). Parental care and mode of fertilisation in ectothermic vertebrates. *Evolution*, **35**, 775–93.

Hahn, M. E. (1983). Genetic 'artifacts' and aggressive behavior. In *Aggressive Behavior: Genetic and Neural Approaches*, ed. E. C. Simmel, M. E. Hahn & J. K. Walters, pp. 67–88. Hillsdale, New Jersey: Erlbaum.

Hall, K. R. L. (1963). Variations in the ecology of the chacma baboon, *Papio ursinus*. *Symposia of the Zoological Society of London*, **10**, 1–28.

Hamilton, W. D. (1964). The genetical evolution of social behaviour, I and II. *Journal of Theoretical Biology*, **7**, 1–52.

Hamilton, W. J. III, Tilson, R. L. & Frank, L. G. (1986). Sexual monomorphism in spotted hyaenas *Crocuta crocuta*. *Ethology*, **71**, 63–73.

Hammerstein, P. (1981). The role of asymmetries in animal conflicts. *Animal Behaviour*, **29**, 193–205.

Hammerstein, P. & Parker, G. A. (1982). The asymmetric war of attrition. *Journal of Theoretical Biology*, **96**, 647–82.

Hannes, R-P., Franck, D. & Liemann, F. (1984). Effects of rank-order fights on whole-body and blood concentrations of androgens and corticosteroids in the male swordtail (*Xiphophorus helleri*). *Zeitschrift für Tierpsychologie*, **65**, 53–65.

Hansen, A. J. & Rohwer, S. (1986). Coverable badges and resource defence in birds. *Animal Behaviour*, **34**, 69–76.

Harding, C. F. (1983). Hormonal influences on avian aggressive behavior. In *Hormones and Aggressive Behavior*, ed. B. B. Svare, pp. 435–67. New York: Plenum.

Harding, C. F. & Leshner, A. I. (1972). The effects of adrenalectomy on the aggressiveness of differentially housed mice. *Physiology and Behavior*, **8**, 437–40.

Harlow, H. F., Dodsworth, R. O. & Harlow, M. K. (1965). Total isolation in monkeys. *Proceedings of the National Academy of Sciences*, **54**, 90–7.

Harvey, I. F. & Corbet, P. S. (1985). Territorial behaviour of larvae enhances mating success of male dragonflies. *Animal Behaviour*, **33**, 561–5.

Harvey, I. F. & Corbet, P. S. (1986). Territorial interactions between larvae of the dragonfly *Pyrrhosoma nymphula:* outcome of encounters. *Animal Behaviour*, **34**, 1550–61.

Harvey, P. H. & Greenwood, P. J. (1978). Anti-predator defence strategies: some evolutionary problems. In *Behavioural Ecology: An Evolutionary Approach*, ed. J. R. Krebs & N. B. Davies, pp. 129–51. Oxford: Blackwell Scientific.

Harvey, P. W. & Chevins, P. F. D. (1985). Crowding pregnant mice affects attack and threat behavior of male offspring. *Hormones and Behavior*, **19**, 86–97.

Hatch, A., Balazs, T., Wiberg, C. S. & Grice, H. C. (1963). Long term isolation in rats. *Science*, **142**, 507.

Hebb, D. O. (1946). On the nature of fear. *Psychological Review*, **53**, 259–76.

Hediger, H. (1950). *Wild Animals in Captivity*. London: Butterworth.

Heiligenberg, W. (1974). Processes governing behavioral states of readiness. In *Advances in the Study of Behavior* 5, ed. D. S. Lehrman, J. S. Rosenblatt, R. A. Hinde & E Shaw, pp. 173–200. New York: Academic Press.

Heller, K. E. (1977). Temporal relationships between shock treatments and intermale fighting behaviour in laboratory mice. Effects of modifying testicular and adrenal functions. *Behavioural Processes*, **2**, 337–47.

Heller, K. E. (1978). Role of corticosterone in the control of post-shock fighting behaviour in male laboratory mice. *Behavioural Processes*, **3**, 211–22.

Heller, K. E. (1979). An attempt to separate the roles of corticosterone and ACTH in the control of post-shock fighting behaviour in male laboratory mice. *Behavioural Processes*, **4**, 231–8.

Henderson, D. L. & Chiszar, D. A. (1977). Analysis of aggressive behaviour in the bluegill sunfish *Lepomis macrochirus rafinesque:* effects of sex and size. *Animal Behaviour*, **25**, 122–30.

Hildrew, A. G. & Townsend, C. R. (1980). Aggregation, interference and foraging by larvae of *Plectrocnemia conspersa* (Trichoptera: Polycentropodidae). *Animal Behaviour*, **28**, 553–60.

Hinde, R. A. (1953). The conflict between drives in the courtship and copulation of the chaffinch. *Behaviour*, **5**, 1–31.

Hinde, R. A. (1956). The biological significance of the territories of birds. *The Ibis*, **98**, 340–69.

Hinde, R. A. (1967). The nature of aggression. *New Society*, **9**, 302–4.

Hinde, R. A. (1970). *Animal Behaviour* (2nd edn). London & New York: McGraw-Hill.

Hinde, R. A. (1975). The concept of function. In *Function and Evolution in Behaviour*, ed. G. P. Baerends, C. Beer & A. Manning, pp. 3–15. Oxford: Clarendon Press.

Hinde, R. A. (1981). Animal signals: ethological and games-theory approaches are not incompatible. *Animal Behaviour*, **29**, 535–42.

Hinde, R. A. (1982). *Ethology*. Oxford: Oxford University Press.

Hixon, M. A., Carpenter, F. L. & Paton, D. C. (1983). Territory area, flower density and time budgeting in hummingbirds: an experimental and theoretical analysis. *American Naturalist*, **122**, 366–91.

Hoar, W. S. (1962). Hormones and the reproductive behaviour of the male three-spined stickleback *Gasterosteus aculeatus*. *Animal Behaviour*, **10**, 247–66.

Hoffman, H. S., Ratner, A. M., Eiserer, L. A. & Grossman, D. J. (1974). Aggressive behavior in immature ducklings. *Journal of Comparative and Physiological Psychology*, **86**, 569–80.

Hogan, J. A. (1967). Fighting and reinforcement in Siamese fighting fish *Betta splendens*. *Journal of Comparative and Physiological Psychology*, **64**, 356–9.

Hogan, J. A. & Roper, T. J. (1978). A comparison of the properties of different reinforcers. In *Advances in the Study of Behavior* 8, ed. J. S. Rosenblatt, R. A. Hinde, C. Beer & M.-C. Busnel, pp. 155–255. New York: Academic Press.

Hole, G. & Einon, D. F. (1984). Play in rodents. In *Play in Animals and Humans*, ed. P. K. Smith, pp. 95–117. Oxford: Blackwell Scientific.

Hood, K. E. (1984). Aggression among female rats during the estrous cycle. In *Biological Perspectives on Aggression*, ed. K. J. Flannelly, R. J. Blanchard & D. C. Blanchard, pp. 181–8. New York: Alan Liss.

Horn, H. S. (1968). The adaptive significance of colonial nesting in Brewer's Blackbird *Euphagus cyancephalus*. *Ecology*, **49**, 682–94.

Horn, H. S. & Rubenstein, D. I. (1984). Behavioural adaptations and life history. In *Behavioural Ecology: An Evolutionary Approach* (2nd edn), ed. J. R. Krebs & N. B. Davies, pp. 279–98. Oxford: Blackwell Scientific.

Horridge, P. A. S. (1970). 'The development of copulatory and fighting behaviour in the domestic chick.' PhD dissertation, University of Sussex, UK.

Houston, A. I., McCleery, R. H. & Davies, N. B. (1985). Territory size, prey renewal and feeding rates: interpretation of observations on the pied wagtail by simulation. *Journal of Animal Ecology*, **54**, 227–39.

Houston, A. I. & McFarland, D. J. (1980). Behavioral resilience and its relation to demand functions. In *Limits to Action: The Allocation of Individual Behavior*, ed. J. E. R. Staddon, pp. 177–203. New York: Academic Press.

Howard, H. E. (1920). *Territory in Bird Life*. London: Murray.

Hrdy, S. B. (1979). Infanticide among animals: a review, classification and examination of the implications for reproductive strategies of individuals. *Ethology and Sociobiology*, **1**, 13–40.

Hrdy, S. B. (1981). *The Woman That Never Evolved*. Cambridge, Massachusetts: Harvard University Press.

Huck, V. W., Bracken, A. C. & Lisk, R. D. (1983). Female-induced pregnancy block in the golden hamster. *Behavioral and Neural Biology*, **38**, 190–3.

Huntingford, F. A. (1976a). The relationship between anti-predator behaviour and aggression among conspecifics in the three-spined stickleback, *Gasterosteus aculeatus*. *Animal Behaviour*, **24**, 245–60.

Huntingford, F. A. (1976b). The relationship between inter- and intra-specific aggression. *Animal Behaviour*, **24**, 485–97.

Huntingford, F. A. (1980). Analysis of the motivational processes underlying aggression in animals. In *Analysis of Motivational Processes*, ed. F. M. Toates & T. R. Halliday, pp. 341–56. London & New York: Academic Press.

Huntingford, F. A. (1984a). *The Study of Animal Behaviour*. London & New York: Chapman & Hall.

Huntingford, F. A. (1984b). Some ethical issues raised by studies of predation and aggression. *Animal Behaviour*, **32**, 210–15.

Huntingford, F. A. & Turner, A. (1987). *Animal Conflict*. London: Chapman & Hall.

Hutchinson, R. R., Renfrew, J. W. & Young, G. A. (1971). Effects of long-term shock and associated stimuli on aggressive and manual responses. *Journal of the Experimental Analysis of Behavior*, **15**, 141–66.

Hutchinson, R. R., Ulrich, R. E. & Azrin, N. H. (1965). Effects of age and related factors on the pain-aggression reaction. *Journal of Comparative and Physiological Psychology*, **59**, 365–9.

Hutzell, R. R. & Knutson, J. F. (1972). A comparison of shock-elicited fighting and shock-elicited biting in rats. *Physiology and Behavior*, **8**, 477–80.

Huxley, J. S. (1934). A natural experiment on the territorial instinct. *British Birds*, **27**, 270–7.

Hyatt, G. W. & Salmon, M. (1978). Combat in the fiddler crabs *Uca pugilator* and *U. pugnax*: a quantitative analysis. *Behaviour*, **65** 182–221.

Hyde, J. S. & Sawyer, T. F. (1977). Estrous cycle fluctuations in aggressiveness of house mice. *Hormones and Behavior*, **9**, 290–5.

Itzkowitz, M. (1985). Sexual differences in offspring defense in a monogamous cichlid fish. *Zeitschrift für Tierpsychologie*, **70**, 247–55.

Jaeger, R. G., Nishikawa, K. C. B. & Barnard, D. E. (1983). Foraging tactics of a terrestrial salamander: costs of territorial defence. *Animal Behaviour*, **31**, 191–8.

Jakobsson, S., Radesater, T. & Jarvi, T. (1979). On the fighting behaviour of *Nanacara anomala* (Pisces, Cichlidae) males. *Zeitschrift für Tierpsychologie*, **49**, 210–20.

Jarman, P. J. (1974). The social organization of the antelope in relation to their ecology. *Behaviour*, **48**, 215–67.

Jenni, D. A. (1974). Evolution of polyandry in birds. *American Zoologist*, **14**, 129–44.

Jennings, H. S. (1906). *Behavior of Lower Organisms*. New York: Columbia University Press.

Johnson, R. N. (1972). *Aggression in Man and Animals*. Philadelphia: Saunders.

Johnson, R. N. & Johnson, L. D. (1973). Intra- and interspecific social and aggressive behaviour in the Siamese fighting fish, *Betta splendens*. *Animal Behaviour*, **21**, 665–72.

Jones, R. B. & Nowell, N. W. (1973). The effect of familiar visual and olfactory cues on the aggressive behavior of mice. *Physiology and Behavior*, **10**, 221–3.

Kagan, J. (1974). Discrepancy, temperament and infant distress. In *The Origins of Fear*, ed. M. Lewis & L. A. Rosenblum, pp. 229–49. New York: Wiley.

Kahn, M. H. (1951). The effects of severe defeat at various age levels on the aggressive behavior of mice. *Journal of Genetic Psychology*, **79**, 117–30.

Katongle, C., Naftolin, F. & Short, R. V. (1971). Relationship between blood levels of luteinizing hormone and testosterone in bulls, and the effects of sexual stimulation. *Journal of Endocrinology*, **50**, 457–60.

Keenleyside, M. H. (1971). Mate desertion in relation to adult sex ratio in the biparental cichlid fish *Herotilapia multispinosa*. *Animal Behaviour*, **31**, 683–8.

Kelleway, L. G. & Brain, P. F. (1982). The utilities of aggression in the viper, *Vipera berus berus* (L.). *Aggressive Behavior*, **8**, 141–3.

King, J. A. (1957). Relationships between early experience and adult aggressive behavior in inbred mice. *Journal of Genetic Psychology*, **90**, 151–66.

King, J. A. & Gurney, N. L. (1954). The effect of early social experience on adult aggressive behavior in *C57Bl/10* mice. *Journal of Comparative and Physiological Psychology*, **47**, 326–30.

Kinsley, C. & Svare, B. B. (1986). Prenatal stress reduces intermale aggression in mice. *Physiology and Behavior*, **36**, 983–6.

Knight, R. L. & Temple, S. A. (1986). Methodological problems in studies of avian nest defence. *Animal Behaviour*, **34**, 561–6.

Knutson, J. F. & Viken, R. J. (1984). Animal analogues of human aggression: studies of social experience and escalation. In *Biological Perspectives on Aggression*, ed. K. J. Flannelly, R. J. Blanchard & D. C. Blanchard, pp. 75–94. New York: Liss.

Kohda, Y. (1983). The effects of color patterns on aggressive behaviors of a freshwater serranid fish, *Coreoperca kawamebari*. *Zoological Magazine*, **92**, 356–60.

Krebs, J. R. (1978). Optimal foraging: decision rules for predators. In *Behavioural Ecology: An Evolutionary Approach*, ed. J. R. Krebs & N. B. Davies, pp. 23–63. Oxford: Blackwell Scientific.

Krebs, J. R. (1982). Territorial defence in the Great Tit (*Parus major*): do residents always win? *Behavioral Ecology and Sociobiology*, **11**, 185–94.

Krebs, J. R. & Davies, N. B. (1981). *An Introduction to Behavioural Ecology*. Oxford: Blackwell Scientific.

Krebs, J. R. & Dawkins, R. (1984). Animal signals: mind reading and manipulaton. In *Behavioural Ecology: An Evolutionary Approach* (2nd edn), ed. J. R. Krebs & N. B. Davies, pp. 380–402. Oxford: Blackwell Scientific.

Krebs, J. R. & McCleery, R. H. (1984). Optimization in behavioural ecology. In *Behavioural Ecology: An Evolutionary Approach* (2nd edn), ed. J. R. Krebs & N. B. Davies, pp. 91–121. Oxford: Blackwell Scientific.

Krsiak, M. & Borgesova, M. (1973). Aggression and timidity induced in mice by isolation. *Activa nervosa supera (Praha)*, **15**, 21–2.

Kruuk, H. (1964). Predators and anti-predator behaviour of the black-headed gull. *Behaviour*, Supplement 11.

Kruuk, H. (1972). *The Spotted Hyena: A Study of Predation and Social Behavior*. Chicago & London: University of Chicago Press.

Kummer, H. (1971). *Primate Societies*. Chicago: Aldine.

Kummer, H., Gotz, W. & Angst, W. (1974). Triadic differentiation: an inhibitory process protecting pair bonds in baboons. *Behaviour*, **49**, 62–87.

Kuo, Z. Y. (1960a). Studies on the basic factors in animal fighting. II. Nutritional factors affecting fighting behavior in quails. *Journal of Genetic Psychology*, **96**, 207–16.

Kuo, Z. Y. (1960b). Studies on the basic factors in animal fighting. III. Hormonal factors affecting fighting in quail. *Journal of Genetic Psychology*, **96**, 217–23.

Kuo, Z. Y. (1960c). Studies on the basic factors in animal fighting. IV. Developmental and environmental factors affecting fighting in quails. *Journal of Genetic Psychology*, **96**, 225–39.

Kuo, Z. Y. (1967). *The Dynamics of Development: An Epigenetic View*. New York: Random House.

Kuse, A. R. & DeFries, J. C. (1976). Social dominance and Darwinian fitness in laboratory mice: an alternative test. *Behavioral Biology*, **16**, 113–16.

Labov, J. B., Huck, U. K., Elwood, R. W. & Brooks, R. J. (1985). Current problems in the study of infanticidal behavior of rodents. *Quarterly Review of Biology*, **60**, 1–20.

Lack, D. (1939). The behaviour of the robin. Parts I and II. *Proceedings of the Zoological Society of London, A,* **109,** 169–219.

Lagerspetz, K. (1964). Studies on the aggressive behaviour of mice. *Annales Academiae Scientiarum Fennicae, Series B,* **131** (3), 7–131.

Lagerspetz, K. & Mettala, R. (1965). Simulation experiments on stimuli eliciting aggressive behaviour in mice. *Reports from the Institute of Psychology, University of Turku, No. 13.*

Lagerspetz, K. & Nurmi, R. (1964). An experiment on the frustration-aggression hypothesis. *Reports from the Institute of Psychology, University of Turku, No. 10.*

Lagerspetz, K. & Wuorinen, K. (1965). Cross fostering experiment with mice selectively bred for aggressiveness and non-aggressiveness. *Reports from the Institute of Psychology, University of Turku, No. 17.*

Lang, J. (1971). Interspecific aggression by scleractinian corals. I. The rediscovery of *Scolymia cubensis* (Milne Edwards & Haime). *Bulletin of Marine Science,* **21,** 952–9.

Lang, J. (1973). Interspecific aggression by scleractinian corals. 2. Why the race is not only to the swift. *Bulletin of Marine Science,* **23,** 260–79.

Larkin, S. & McFarland, D. J. (1978). The cost of changing from one activity to another. *Animal Behaviour,* **26,** 1237–46.

Lazarus, J. & Crook, J. H. (1973). The effects of luteinizing hormone, oestrogen and ovariectomy on the agonistic behaviour of female *Quelea quelea. Animal Behaviour,* **21,** 49–60.

Lea, R. W., Vowles, D. M. & Dick, H. R. (1986). Factors affecting prolactin secretion during the breeding cycle of the ring dove (*Streptopelia risoria*) and its possible role in incubation. *Journal of Endocrinology,* **110,** 477–58.

Lea, S. E. G. (1978). The psychology and economics of demand. *Psychological Bulletin,* **85,** 441–6.

Le Boeuf, B. J. (1974). Male–male competition and reproductive success in elephant seals. *American Zoologist,* **14,** 163–76.

Legrand, R. (1970). Successful aggression as the reinforcer for runway behavior of mice. *Psychonomic Science,* **20,** 303–5.

Legrand, R. & Fielder, R. (1973). Role of dominance submission relationships in shock-induced fighting of mice. *Journal of Comparative and Physiological Psychology,* **82,** 501–6.

Lemmetyinen, R. (1971). Nest defence behaviour of common and arctic terns and its effects on the success achieved by predators. *Ornis Fennica,* **48,** 13–24.

Lemmetyinen, R. (1972). Nest defence behaviour in the arctic tern, *Sterna paradisaea,* towards stuffed nest predators on Spitzbergen. *Reports of Kevo Subarctic Research Station,* **9,** 28–31.

Leshner, A. I. (1975). A model of hormones and agonistic behavior. *Physiology and Behavior,* **15,** 225–35.

Leshner, A. I. (1980). The interaction of experience and neuroendocrine factors in determining behavioral adaptations to aggression. *Progress in Brain Research,* **53,** 427–38.

Leshner, A. I. (1981). The role of hormones in the control of submissiveness. In *A Multidisciplinary Approach to Aggression Research,* ed. P. F. Brain & D. Benton, pp. 309–22. Amsterdam: Elsevier/North Holland.

Leshner, A. I. (1983*a*). Pituitary–adrenocortical effects on intermale agonistic behavior. In *Hormones and Aggressive Behavior*, ed. B. B. Svare, pp. 27–37. New York: Plenum.

Leshner, A. I. (1983*b*). The hormonal responses to competition and their behavioral significance. In *Hormones and Aggressive Behavior*, ed. B. B. Svare, pp. 393–404. New York: Plenum.

Leshner, A. I. & Moyer, J. A. (1975). Androgens and agonistic behavior in mice: relevance to aggression and irrelevance to avoidance of attack. *Physiology and Behavior*, **15**, 695–9.

Leshner, A. I., Moyer, J. A. & Walker, W. A. (1975). Pituitary–adrenocortical activity and avoidance-of-attack in mice. *Physiology and Behavior*, **15**, 689–93.

Lewontin, R. C. (1979). Fitness, survival and optimality. In *Analysis of Ecological Systems*, ed. D. J. Horn, G. R. Stairs & R. D. Mitchell, pp. 3–21. Columbus: Ohio State University.

Leyhausen, P. (1956). Verhaltensstudien bei Katzen. *Zeitschrift für Tierpsychologie, Beiheft 2*.

Leyhausen, P. (1973). On the functions of the relative hierarchy of moods. In *Motivation of Human and Animal Behavior: An Ethological View*, ed. K. Lorenz & P. Leyhausen, pp. 144–247. New York: Van Nostrand.

Leyhausen, P. (1979). *Cat Behavior*. New York: Garland STPM.

Lincoln, G. A., Guinness, F. & Short, R. V. (1972). The way in which testosterone controls the social and sexual behavior of the red deer stag *Cervus elaphus*. *Hormones and Behavior*, **3**, 375–96.

Lisk, R. D., Ciaccio, L. A. & Catanzaro, C. (1983). Mating behaviour of the golden hamster under seminatural conditions. *Animal Behaviour*, **31**, 659–66.

Lloyd, J. A. & Christian, J. J. (1967). Relationship of activity and aggression to density in two confined populations of house mice *Mus musculus*. *Journal of Mammalogy*, **48**, 262–9.

Loher, W. & Huber, F. (1966). Nervous and endocrine control of sexual behaviour in a grasshopper (*Gomphocerus rufus* L., Acridinae). *Symposia of the Society for Experimental Biology and Medicine*, **20**, 381–400.

Looney, T. A. & Cohen, P. S. (1982). Aggression induced by intermittent positive reinforcement. *Biobehavioral Reviews*, **6**, 15–37.

Lore, R. K. & Stipo-Flaherty, A. (1984). Postweaning social experience and adult aggression in rats. *Physiology and Behavior*, **33**, 571–4.

Lore, R. & Takahashi, L. (1984). Postnatal influences on intermale aggression in rodents. In *Biological Perspectives on Aggression*, ed. K. J. Flannelly, R. J. Blanchard & D. C. Blanchard, pp. 189–206. New York: Liss.

Lorenz, K. (1950). The comparative method in studying innate behaviour patterns. *Symposia of the Society for Experimental Biology*, **4**, 221–68.

Lorenz, K. (1958). The evolution of behavior. *Scientific American*, **199** (6), 67–78.

Lorenz, K. (1966). *On Aggression*. New York: Harcourt, Brace & World.

Louch, C. D. (1956). Adrenocortical activity in relation to the density and dynamics of three confined populations of *Microtus*. *Ecology*, **37**, 701–13.

Loy, J. (1970). Behavioral responses of free-ranging rhesus monkeys to food storage. *American Journal of Physical Anthropology*, **33**, 263–71.

Lumia, A. R., Rieder, C. A. & Reynierse, J. H. (1973). The differential effects of reinforcement and testosterone on aggressive responding in pigeons: species typical and aversive aspects of pigeon aggression. *Bulletin of the Psychonomic Society*, **1**, 165–6.

Lyon, D. O. & Ozolins, D. (1970). Pavlovian conditioning of shock-elicited aggression: a discrimination procedure. *Journal of the Experimental Analysis of Behavior*, **13**, 325–31.

MacArthur, R. H. (1972). *Geographical Ecology: Patterns in the Distribution of Species*. New York: Harper & Row.

McCann, T. (1982). Aggressive and maternal activities of female southern elephant seals (*Mirounga leonina*). *Animal Behaviour*, **30**, 268–76.

McCarty, R. & Southwick, C. H. (1979). Parental environment: effects on survival, growth and aggressive behaviors of two rodent species. *Developmental Psychobiology*, **12**, 269–79.

McClearn, G. E. & DeFries, J. C. (1973). Genetics and mouse aggression. In *The Control of Aggression*, ed. J. F. Knutson, pp. 59–77. Chicago: Aldine.

McCleery, R. H. (1978). Optimal behaviour sequences and decision making. In *Behavioural Ecology: An Evolutionary Approach*, ed. J. R. Krebs & N. B. Davies, pp. 377–410. Oxford: Blackwell Scientific.

McFarland, D. J. (1966). On the causal and functional significance of displacement activities. *Zeitschrift für Tierpsychologie*, **23**, 217–35.

McFarland, D. J. (1969). Mechanisms of behavioural disinhibition. *Animal Behaviour*, **17**, 238–42.

McFarland, D. J. (1976). Form and function in the temporal organisation of behaviour. In *Growing Points in Ethology*, ed. P. P. G. Bateson & R. A. Hinde, pp. 55–93. Cambridge: Cambridge University Press.

McFarland, D. J. (1977). Decision making in animals. *Nature*, **269**, 15–21.

McFarland, D. J. (1985). *Animal Behaviour*. London: Pitman.

Mackintosh, J. H. (1970). Territory formation by laboratory mice. *Animal Behaviour*, **18**, 177–83.

Mackintosh, J. H. (1973). Factors affecting the recognition of territory boundaries by mice (*Mus musculus*). *Animal Behaviour*, **21**, 464–70.

Mackintosh, J. H. & Grant, E. C. (1966). The effect of olfactory stimuli on the agonistic behaviour of laboratory mice. *Zeitschrift für Tierpsychologie*, **23**, 584–7.

McLean, I. G. (1983). Paternal behaviour and killing of young in arctic ground squirrels. *Animal Behaviour*, **31**, 32–44.

Mallory, F. F. & Brooks, R. J. (1978). Infanticide and other reproductive strategies in the collared lemming *Dicrostonyx groenlandicus*. *Nature*, **273**, 144–6.

Manzur, M. I. & Fuentes, E. R. (1979). Polygyny and agonistic behavior in the tree-dwelling lizard *Liolaemus tenius* (Iguanidae). *Behavioral Ecology and Sociobiology*, **6**, 23–8.

Marler, P. (1956*a*). Studies of proximity in chaffinches. (3) Proximity as a cause of aggression. *Animal Behaviour*, **4**, 23–30.

Marler, P. (1956*b*). Territory and individual distance in the chaffinch *Fringilla coelebs*. *The Ibis*, **98**, 496–501.

Marler, P. (1976). On animal aggression: the roles of strangeness and familiarity. *American Psychologist*, **31**, 239–46.

Marques, D. M. & Valenstein, E. S. (1977). Individual differences in aggressiveness of female hamsters: response to intact and castrated males and to females. *Animal Behaviour*, **25**, 131–9.

Martin, P. & Caro, T. (1985). On the functions of play and its role in behavioral development. In *Advances in the Study of Behavior*, vol. 15, ed. J. S. Rosenblatt, C. Beer, M.-C. Busnel & P. J. B. Slater, pp. 59–103. New York: Academic Press.

Mason, W. A. (1960). The effect of social restriction on the behavior of rhesus monkeys. 1. Free social behavior. *Journal of Comparative and Physiological Psychology*, **53**, 582–9.

Mason, W. A. (1963). The effects of environmental restriction on the social development of rhesus monkeys. In *Primate Social Behavior*, ed. C. H. Southwick, pp. 161–73. Princeton: Van Nostrand.

Matthewson, S. (1961). Gonadotrophic hormones affect aggressive behavior in starlings. *Science*, **134**, 1522–3.

Maynard Smith, J. (1964). Group selection and kin selection. *Nature*, **210**, 1145–7.

Maynard Smith, J. (1972). *On Evolution*. Edinburgh: Edinburgh University Press.

Maynard Smith, J. (1974). The theory of games and the evolution of animal conflicts. *Journal of Theoretical Biology*, **47**, 209–21.

Maynard Smith, J. (1976). Evolution and the theory of games. *American Scientist*, **64**, 41–5.

Maynard Smith, J. (1977). Parental investment: a prospective analysis. *Animal Behaviour*, **25**, 1–9.

Maynard Smith, J. (1978). Optimization theory in evolution. *Annual Review of Ecology and Systematics*, **9**, 31–56.

Maynard Smith, J. (1982). *Evolution and the Theory of Games*. Cambridge & New York: Cambridge University Press.

Maynard Smith, J. & Parker, G. A. (1976). The logic of asymmetric contests. *Animal Behaviour*, **24**, 159–75.

Maynard Smith, J. & Price, G. R. (1973). The logic of animal conflict. *Nature*, **246**, 15–18.

Maynard-Smith, J. & Riechert, S. E. (1984). A conflicting-tendency model of spider agonistic behaviour: hybrid–pure population line comparisons. *Animal Behaviour*, **32**, 564–78.

Mayo, O. (1983). *Natural Selection and its Constraints*. London & New York: Academic Press.

Mayr, E. (1982). Teleological and teleonomic: a new analysis. In *Learning, Development and Culture*, ed. H. C. Plotkin, pp. 17–38. New York: Wiley.

Mazur, A. & Lamb, T. A. (1980). Testosterone, status and mood in human males. *Hormones and Behavior*, **14**, 236–46.

Meaney, M. J. & Stewart, J. (1981). A descriptive study of social development in the rat (*Rattus norvegicus*). *Animal Behaviour*, **29**, 34–45.

Melvin, K. B. (1985). Attack/display as a reinforcer in *Betta splendens*. *Bulletin of the Psychonomic Society*, **23**, 350–2.

Miller, N. E. (1941). The frustration-aggression hypothesis. *Psychological Review*, **48**, 337–42.

Miller, N. E. (1979). Dollard, John. In *International Encyclopaedia of the Social Sciences*, vol. 18, ed. D. L. Sills. New York: The Free Press.

Miller, N. E. & Dollard, J. (1941). *Social Learning and Imitation*. New Haven: Yale University Press.

Miller, P. H. (1983). *Theories of Developmental Psychology*. San Francisco: Freeman.

Milligan, W. L., Powell, D. A. & Borasio, G. (1973). Sexual variables and shock-elicited aggression. *Journal of Comparative and Physiological Psychology*, **83**, 441–50.

Mitchell, K. A. (1976). Competitive fighting for shells in the hermit crab, *Clibanarius vittatus*. *Aggressive Behavior*, **2**, 31–7.

Montgomery, E. L. & Caldwell, R. E. (1984). Aggressive brood defense by females in the stomatopod *Gonodactylus bredini*. *Behavioral Ecology and Sociobiology*, **14**, 247–51.

Morris, D. (1957). 'Typical intensity' and its relation to the problem of ritualisation. *Behaviour*, **11**, 1–12.

Mosler, H.-J. (1985). Making the decision to continue the fight or to flee: an analysis of contests between male *Haplochromis burtoni* (Pisces). *Behaviour*, **92**,129–45.

Moss, R., Kolb, H. H., Marquiss, M., Watson, A., Treca, B., Watt, D. & Glennie, W. (1979). Aggressiveness and dominance in captive cock red grouse. *Aggressive Behavior*, **5**, 59–84.

Moyer, K. E. (1968). Kinds of aggression and their physiological basis. *Communications in Behavioral Biology A*, **2**, 65–87.

Mugford, R. A. (1973). Intermale fighting affected by home-cage odors of male and female mice. *Journal of Comparative and Physiological Psychology*, **84**, 289–95.

Mugford, R. A. & Nowell, N. W. (1971). The preputial glands as a source of aggression-promoting odors in mice. *Physiology and Behavior*, **6**, 247–9.

Mugford, R. A. & Nowell, N. W. (1972). Paternal stimulation during infancy: effects upon aggression and open-field performance of mice. *Journal of Comparative and Physiological Psychology*, **79**, 30–6.

Munro, A. D. & Pitcher, T. J. (1985). Steroid hormones and agonistic behavior in a cichlid teleost *Aequidens pulcher*. *Hormones and Behavior*, **19**, 353–71.

Mykytowycz, R. (1961). Social behaviour of an experimental colony of wild rabbits. IV Conclusion: Outbreak of myxomatosis, third breeding season and starvation. *CSIRO Wildlife Research*, **6**, 142–55.

Mykytowycz, R. (1962). Territorial function of chin-gland secretion in the rabbit. *Nature*, **193**, 799.

Mykytowycz, R. (1966). Observations on odoriferous and other glands in the Australian wild rabbit, *Oryctolagus cuniculus* (L.), and the hare, *Lepus europaeus* P. I. The anal gland. *CSIRO Wildlife Research*, **11**, 11–29.

Namikas, J. & Wehmer, F. (1978). Gender composition of the litter affects behavior of male mice. *Behavioral Biology*, **23**, 219–24.

Neil, S. J. (1985). Size assessment and cues: studies of hermit crab contests. *Behaviour*, **92**, 22–38.

Nicholson, A. J. (1954). An outline of the dynamics of animal populations. *Australian Journal of Zoology*, **2**, 9–65.

Noirot, E., Goyens, J. & Buhot, M.-C. (1975). Aggressive behavior of pregnant mice towards males. *Hormones and Behavior*, **6**, 9–17.

Norman, R. F., Taylor, P. D. & Robertson, R. J. (1977). Stable equilibrium strategies and penalty functions in a game of attrition. *Journal of Theoretical Biology*, **65**, 571–8.

O'Keefe, J. & Nadel, L. (1979). Precis of O'Keefe and Nadel's 'The hippocampus as a cognitive map'. *The Behavioral and Brain Sciences*, **2**, 487–533.

O'Kelly, L. E. & Steckle, L. C. (1939). A note on long-enduring emotional responses in the rat. *Journal of Psychology*, **8**, 125–31.

O'Neill, K. M. (1983). The significance of body size in territorial interactions of male beewolves (Hymenoptera: Sphecidae, *Philanthus*). *Animal Behaviour*, **31**, 404–11.

Olomon, C. M., Breed, M. D. & Bell, W. J. (1976). Ontogenetic and temporal aspects of agonistic behavior in a cockroach, *Periplaneta americana*. *Behavioral Biology*, **17**, 243–8.

Olweus, D. (1979). Stability of aggressive reactions in males: a review. *Psychological Bulletin*, **86**, 852–75.

Olweus, D. (1984). Development of stable aggressive reaction patterns in males. In *Advances in the Study of Aggression*, vol. 1, ed. R. J. Blanchard & D. C. Blanchard, pp. 103–37. New York: Academic Press.

Ostermeyer, M. C. (1983). Maternal aggression. In *Parental Behaviour of Rodents*, ed. R. W. Elwood, pp. 151–79. Chichester: Wiley.

Owings, D. H., Borchert, M. & Virginia, R. (1977). The behaviour of California ground squirrels. *Animal Behaviour*, **25**, 221–30.

Owings, D. H. & Loughry, W. J. (1985). Variation in snake-elicited jump-yipping by black-tailed Prairie dogs: ontogeny and snake-specificity. *Zeitschrift für Tierpsychologie*, **70**, 177–200.

Packer, C. (1977). Reciprocal altruism in *Papio anubis*. *Nature*, **265**, 441–3.

Parker, G. A. (1970). Sperm competition and its evolutionary consequences in the insects. *Biological Reviews*, **45**, 525–67.

Parker, G. A. (1974*a*). Courtship persistence and female-guarding as male time investment strategies. *Behaviour*, **48**, 157–84.

Parker, G. A. (1974*b*). Assessment strategy and the evolution of fighting behavior. *Journal of Theoretical Biology*, **47**, 223–43.

Parker, G. A. & Rubenstein, D. I. (1981). Role assessment, reserve strategy, and acquisition of information in asymmetric animal conflicts. *Animal Behaviour*, **29**, 221–40.

Parker, G. A. & Thompson, E. A. (1980). Dung fly struggles: a test of the war of attrition. *Behavioral Ecology and Sociobiology*, **7**, 37–44.

Payne, A. P. & Swanson, H. H. (1970). Agonistic behaviour between pairs of hamsters of the same and opposite sex in a neutral observation area. *Behaviour*, **36**, 259–69.

Payne, A. P. & Swanson, H. H. (1971). Hormonal modification of aggressive behaviour between female golden hamsters. *Journal of Endocrinology*, **51**, xvii–xviii.

Payne, A. P. & Swanson, H. H. (1972). The effect of sex hormones on the aggressive behaviour of the female golden hamster, *Mesocricetus auratus* Waterhouse. *Animal Behaviour*, **20**, 782–7.

Peeke, H. V. S. (1969). Habituation of conspecific aggression in the three-spine stickleback *Gasterosteus aculeatus*. *Behaviour*, **35**, 137–56.

Peeke, H. V. S. (1982). Stimulus- and motivation-specific sensitization and redirection of aggression in the three-spined stickleback (*Gasterosteus aculeatus*). *Journal of Comparative and Physiological Psychology*, **96**, 816–22.

Peeke, H. V. S., Herz, M. J. & Gallagher, J. E. (1971). Changes in aggressive interaction in adjacently territorial convict cichlids *Cichlasoma nigrofasciatum*: a study of habituation. *Behaviour*, **40**, 43–54.

Peeke, H. V. S. & Peeke, S. C. (1970). Habituation of aggressive responses in the Siamese fighting fish. *Behaviour*, **36**, 232–45.

Peeke, H. V. S. & Peeke, S. C. (1982). Parental factors in the sensitization and habituation of territorial aggression in the Convict Cichlid (*Cichlasoma nigrofasciatum*). *Journal of Comparative and Physiological Psychology*, **96**, 955–66.

Peeke, H. V. S. & Veno, A. (1973). Stimulus specificity of habituated aggression in the stickleback *Gasterosteus aculeatus*. *Behavioral Biology*, **8**, 427–32.

Peeke, H. V. S. & Veno, A. (1976). Response independent habituation of territorial aggression in the three-spined stickleback (*Gasterosteus aculeatus*). *Zeitschrift für Tierpsychologie*, **40**, 53–8.

Peeke, H. V. S., Wyers, E. J. & Herz, M. J. (1969). Waning of the aggressive response to male models in the three-spined stickleback (*Gasterosteus aculeatus* L.). *Animal Behaviour*, **17**, 224–8.

Petrie, M. (1984). Territory size in the moorhen (*Gallinula chloropus*): an outcome of RHP asymmetry between neighbours. *Animal Behaviour*, **32**, 861–70.

Plotkin, H. C. & Odling-Smee, F. J. (1979). Learning, change and evolution: an enquiry into the teleonomy of learning. In *Advances in the Study of Behavior*, 10, ed. J. S. Rosenblatt, R. A. Hinde, C. Beer & M.-C. Busnel, pp. 1–42. New York: Academic Press.

Politch, J. A. & Herrenkohl, L. R. (1979). Prenatal stress reduces maternal aggression by mice offspring. *Physiology and Behavior*, **23**, 415–18.

Poole, T. B. (1974). Detailed analysis of fighting in polecats (*Mustelidae*) using cine film. *Journal of Zoology*, **173**, 369–93.

Poole, T. B. & Fish, J. (1975). An investigation of playful behaviour in *Rattus norvegicus* and *Mus musculus* (Mammalia). *Journal of Zoology*, **175**, 61–71.

Porter, R. H. (1976). Sex differences in the agonistic behaviour of spiny mice. *Zeitschrift für Tierpsychologie*, **40**, 100–8.

Potegal, M. (1979). The reinforcing value of several types of aggressive behavior: a review. *Aggressive Behavior*, **5**, 353–73.

Potegal, M. & tenBrink, L. (1984). Behavior of attack-primed and attack-satiated female golden hamsters. (*Mesocricetus auratus*). *Journal of Comparative Psychology*, **98**, 66–75.

Powers, W. T. (1973). Feedback: beyond behaviorism. *Science*, **179**, 351–6.

Powers, W. T. (1978). Quantitative analysis of purposive systems: some spadework at the foundations of scientific psychology. *Psychological Review*, **85**, 417–35.

Puckett, K. J. & Dill, L. M. (1985). The energetics of feeding territoriality in juvenile coho salmon *Oncorhynchus kisutch*. *Behaviour*, **92**, 97–111.

Purvis, K. & Haynes, N. B. (1974). Short-term effects of copulation, human chorionic gonadotrophin injection, and non-tactile association with a female on testosterone levels in the male rat. *Journal of Endocrinology*, **60**, 429–39.

Pyke, G. H. (1979). The economics of territory size and time budget in the golden-winged sunbird. *American Naturalist*, **114**, 131–45.

Rajecki, D. W., Ivins, B. & Rein, B. (1976). Social discrimination and aggressive pecking in domestic chicks. *Journal of Comparative and Physiological Psychology*, **90**, 442–52.

Rand, W. M. & Rand, A. S. (1976). Agonistic behaviour in nesting iguanas: a stochastic analysis of dispute settlement dominated by minimization of energy cost. *Zeitschrift für Tierpsychologie*, **40**, 279–99.

Rasa, O. A. E. (1969). Territoriality and the establishment of dominance by means of visual cues in *Pomacentrus jenkinsi* (Pisces: Pomacentridae). *Zeitschrift für Tierpsychologie*, **26**, 825–45.

Rasa, O. A. E. (1976). Aggression: appetite or aversion? An ethologist's viewpoint. *Aggressive Behavior*, **2**, 213–22.

Rasa, O. A. E. (1981). Ethological aspects of aggression in subhuman animals. In *The Biology of Aggression*, ed. P. F. Brain & D. Benton, pp. 585–601. Rockville, Maryland: Sijthoff & Nordhoff.

Regelmann, K. & Curio, E. (1986). Why do great tit (*Parus major*) males defend their brood more than females do? *Animal Behaviour*, **34**, 1206–14.

Reiter, J., Panken, K. J. & Le Bouef, B. J. (1981). Female competition and reproductive success in northern elephant seals. *Animal Behaviour*, **29**, 670–87.

Reynolds, G. S., Catania, A. C. & Skinner, B. F. (1963). Conditioned and unconditioned aggression in pigeons. *Journal of the Experimental Analysis of Behavior*, **6**, 73–4.

Richards, S. M. (1974). The concept of dominance and methods of assessment. *Animal Behaviour*, **22**, 914–30.

Ridley, M. (1978). Paternal care. *Animal Behavior*, **26**, 904–32.

Riechert, S. E. (1978). Games spiders play: behavioral variability in territorial disputes. *Behavioral Ecology and Sociobiology*, **3**, 135–62.

Riechert, S. E. (1979). Games spiders play II: resource assessment strategies. *Behavioral Ecology and Sociobiology*, **6**, 121–8.

Riechert, S. E. (1984). Games spiders play III: cues underlying context-associated changes in agonistic behaviour. *Animal Behaviour*, **32**, 1–15.

Robertson, J. G. M. (1986). Male territoriality, fighting and assessment of fighting ability in the Australian frog, *Uperoleia rugosa*. *Animal Behaviour*, **34**, 763–72.

Robertson, R. J. & Bierman, G. C. (1979). Parental investment strategies determined by expected benefits. *Zeitschrift für Tierpsychologie*, **50**, 124–8.

Robinson, S. K. (1986). Fighting and assessment in the yellow-rumped cacique (*Cacicus cela*). *Behavioral Ecology and Sociobiology*, **18**, 39–44.

Robitaille, J. A. & Bovet, J. (1976). Field observations on the social behaviour of the Norway rat, *Rattus norvegicus* (Berkendhout). *Biology of Behaviour*, **1**, 289–308.

Rodgers, R. J. (1981a). Drugs, aggression and behavioural methods. In *Multidisciplinary Approaches to Aggression Research*, ed. P. F. Brain & D. Benton, pp. 325–40. Amsterdam: Elsevier/North Holland.

Rodgers, R. J. (1981b). Pain and aggression. In *The Biology of Aggression*, ed. P. F. Brain & D. Benton, pp. 519–27. Rockville, Maryland: Sijthoff & Noordhoff.

Rohwer, S. (1985). Dyed birds achieve higher social status than controls in Harris' sparrows. *Animal Behaviour*, **33**, 1325–31.

Rohwer, S. & Rohwer, F. C. (1978). Status signalling in Harris sparrows: experimental deceptions achieved. *Animal Behaviour*, **26**, 1012–22.

Roper, T. J. (1981). What is meant by the term 'schedule-induced', and how general is schedule induction? *Animal Learning and Behavior*, **9**, 433–40.

Roper, T. J. & Crossland, G. (1982). Mechanisms underlying eating–drinking transitions in rats. *Animal Behaviour*, **30**, 602–14.

Rose, R. M., Gordon, T. P. & Bernstein, I. S. (1972). Plasma testosterone levels in the male rhesus: influences of sexual and social stimuli. *Science*, **178**, 643–5.

Roseler, P.-F., Roseler, I., Strambi, A. & Augier, R. (1984). Influence of insect hormones on the establishment of dominance hierarchies among foundresses of the paper wasp, *Polistes gallicus*. *Behavioral Ecology and Sociobiology*, **15**, 133–42.

Roseler, P.-F., Roseler, I. & Strambi, A. (1986). Role of ovaries and ecdysteroid in dominance hierarchy establishment among foundresses of the primitively social wasp, *Polistes gallicus*. *Behavioral Ecology and Sociobiology*, **18**, 9–13.

Rowell, T. E. (1974). The concept of social dominance. *Behavioral Biology*, **11**, 131–54.

Rowland, W. J. (1982). The effects of male nuptial coloration on stickleback aggression: a re-examination. *Behaviour*, **80**, 118–26.

Rowland, W. J. & Sevenster, P. (1985). Sign stimuli in the threespine stickleback (*Gasterosteus aculeatus*): a re-examination and extension of some classic experiments. *Behaviour*, **93**, 241–57.

Rubenstein, D. I. (1981). Population density, resource patterning, and territoriality in the Everglades pygmy sunfish. *Animal Behaviour*, **29**, 155–72.

Runfeldt, S. & Wingfield, J. C. (1985). Experimentally prolonged sexual activity in female sparrows delays termination of reproductive activity in their untreated mates. *Animal Behaviour*, **33**, 403–10.

Saito, Y. (1986a). Biparental defence in a spider mite (Acari: Tetranychidae) infesting *Sasa* bamboo. *Behavioral Ecology and Sociobiology*, **18**, 377–86.

Saito, Y. (1986b). Prey kills predator: counter-attack success of a spider mite against its specific phytoseiid predator. *Experimental and Applied Acarology*, **2**, 47–62.

Sakagami, S. F. (1954). Occurrence of an aggressive behaviour in queenless hives, with considerations on the social organization of honeybee. *Insectes Sociaux*, **1**, 331–43.

Salzen, E. A. (1962). Imprinting and fear. *Symposium of the Zoological Society of London*, **8**, 199–217.

Schaller, G. B. & Emlen, J. T. (1962). The ontogeny of avoidance behaviour in some precocial birds. *Animal Behaviour*, **10**, 370–81.

Schoener, T. W. (1983). Simple models of optimal territory size, a reconciliation. *American Naturalist*, **121**, 608–29.

Schuurman, T. (1980). Hormonal correlates of agonistic behavior in adult male rats. *Progress in Brain Research*, **53**, 415–20.

Schwagmeyer, P. L. & Woontner, S. J. (1986). Scramble competition polygyny in thirteen-lined ground squirrels: the relative contributions of overt conflict and competitive mate searching. *Behavioral Ecology and Sociobiology*, **19**, 359–64.

Scott, J. P. (1944). Social behavior, range and territoriality in domestic mice. *Proceedings of the Indiana Academy of Sciences*, **53**, 188–95.

Scott, J. P. (1946). Incomplete adjustment caused by frustration of untrained fighting mice. *Journal of Comparative Psychology*, **39**, 379–90.

Scott, J. P. & Fredericson, E. (1951). The causes of fighting in mice and rats. *Physiological Zoology*, **24**, 273–309.

Selander, R. K. (1972). Sexual selection and dimorphism in birds. In *Sexual Selection and the Descent of Man*, ed. B. Campbell, pp. 180–230. Chicago: Aldine.

Sevenster-Bol, A. C. A. (1962). On the causation of drive reduction after a consummatory act. *Archives Neerlandaises de Zoologie*, *Tome 15*, **2**, 174–236.

Sherman, P. A. (1981). Reproductive competition and infanticide in Belding's ground squirrels and other animals. In *Natural Selection and Social Behavior: Recent Research and New Theory*, ed. R. D. Alexander & D. W. Tickle, pp. 311–31. New York: Chiron.

Short, R. V. (1980). The origins of human sexuality. In *Reproduction in Mammals*, vol. 8, *Human Sexuality*, ed. C. R. Austin & R. V. Short, pp. 1–33. Cambridge: Cambridge University Press.

Siann, G. (1985). *Accounting for Aggression*. London & Boston: Allen & Unwin.

Sigg, E. B., Day, C. A. & Colombo, C. (1966). Endocrine factors in isolation-induced aggressiveness in rodents. *Endocrinology*, **78**, 679–84.

Sigg, H. & Falett, J. (1985). Experiments on respect of possession and property in hamadryas baboons (*Papio hamadryas*). *Animal Behaviour*, **33**, 978–84.

Sigurjonsdottir, H. & Parker, G. A. (1981). Dung fly struggles: evidence for assessment strategy. *Behavioral Ecology and Sociobiology*, **8**, 219–30.

Simpson, M. J. A. (1968). The display of the Siamese fighting fish *Betta splendens*. *Animal Behaviour Monographs*, **1**, 1–73.

Sinclair, M. E. (1977). Agonistic behaviour of the stone crab, *Menippe mercenaria* (say). *Animal Behaviour*, **25**, 193–207.

Sivinski, J. (1978). Intrasexual aggression in the stick insects *Diapheromera veliei* and *D. covilleae* and sexual dimorphism in the Phasmatodea. *Psyche*, **85**, 395–405.

Slater, P. J. B. (1978). *Sex Hormones and Behaviour*. (Studies in Biology no. 103.) London: Edward Arnold.

Sluckin, W. (1972). *Imprinting and Early Learning* (2nd edn). London: Methuen.

Smith, A. T., & Ivins, B. L. (1986). Territorial intrusions by pikas (*Ochotona princeps*) as a function of occupant activity. *Animal Behaviour*, **34**, 392–7.

Southwick, C. H. (1967). Effect of maternal environment on aggressive behavior of inbred mice. *American Zoologist*, **7**, 794.

Southwick, C. H. (1968). Effect of maternal environment on aggressive behavior of inbred mice. *Communications in Behavioral Biology*, A, **1**, 129–32.

Southwick, C. H. (1969). Aggressive behaviour of rhesus monkeys in natural and captive groups. In *Aggressive Behaviour*, ed. S. Garattini & E. B. Sigg, pp. 32–43. Amsterdam: Excerpta Medica Foundation.

Sparks, J. (1982). *The Discovery of Animal Behaviour*. London: Collins/BBC.

Spinage, G. A. (1969). Territoriality and social organization of the Uganda defassa waterbuck, *Kobus defassa ugandae*. *Journal of Zoology*, **159**, 329–61.

Staddon, J. E. R. (1984). *Adaptive Behavior and Learning*. New York: Cambridge University Press.

Stamps, J. (1983). Territoriality and the defence of predator-refuges in juvenile lizards. *Animal Behaviour*, **31**, 857–70.

Steiniger, F. (1950). Beitrage zür soziologie und sonstigen biologie der Wanderatte. *Zeitschrift für Tierpsychologie*, **7**, 356–79.

Stephens, M. L. (1982). Mate takeover and possible infanticide by a female Northern Jacana (*Jacana spinosa*). *Animal Behaviour*, **30**, 1252–61.

Stimson, J. (1970). Territorial behaviour in the owl limpet, *Lottia gigantea*. *Ecology*, **51**, 113–18.

Stinson, C. H. (1979). On the selective advantage of fraticide in raptors. *Evolution*, **33**, 1219–25.

Stokes, A. W. (1962). Agonistic behaviour among Blue Tits at a winter feeding station. *Behaviour*, **19**, 118–38.

Storr, A. (1968). *Human Aggression*. Harmondsworth, Middlesex: Allen Lane.

Studd, M. V. & Robertson, R. J. (1985). Evidence for reliable badges of status in territorial yellow warblers (*Dendroica petechia*). *Animal Behaviour*, **33**, 1102–13.

Suarez, S. D. & Gallup, G. G. Jr (1983). Emotionality and fear in birds: a selected review and reinterpretation. *Bird Behaviour*, **5**, 22–30.

Suter, R. B. & Keiley, M. (1984). Agonistic interactions between male *Frontinella pyramitela* (Areneae, Linyphiidae). *Behavioral Ecology and Sociobiology*, **15**, 1–7.

Svare, B. B. (1981). Maternal aggression in mammals. In *Parental Care in Mammals*, ed. D. J. Gubernick & P. H. Klopfer, pp. 179–210. New York: Plenum.

Svare, B. B. (1983). Psychobiological determinants of maternal aggressive behavior. In *Aggressive Behavior: Genetic and Neural Approaches*, ed. E. C. Simmel, M. E. Hahn & J. K. Walters, pp. 129–46. Hillsdale, New Jersey: Lawrence Erlbaum.

Svare, B. B., Betteridge, C., Katz, D. & Samuels, O. (1981). Some situational and experiential determinants of maternal aggression in mice. *Physiology and Behavior*, **26**, 253–8.

Svare, B. B. & Gandelman, R. (1973). Postpartum aggression in mice: experiental and environmental factors. *Hormones and Behavior*, **4**, 323–34.

Svare, B. B. & Mann, M. A. (1983). Hormonal influences on maternal aggression. In *Hormones and Aggressive Behavior*, ed. B. B. Svare, pp. 91–104. New York: Plenum.

Svare, B. B., Miele, J. & Kinsley, C. (1986). Mice: progesterone stimulates aggression in pregnancy-terminated females. *Hormones and Behavior*, **20**, 194–200.

Syme, G. J. (1974). Competitive orders as measures of social dominance. *Animal Behaviour*, **22**, 931–40.

Takahashi, L. K. & Lisk, R. D. (1983). Organization and expression of agonistic and sociosexual behavior in golden hamsters over the estrous cycle and after ovariectomy. *Physiology and Behavior*, **31**, 477–82.

Takahashi, L. K. & Lisk, R. D. (1984). Intrasexual interactions among female golden hamsters (*Mesocricetus auratus*) over the estrous cycle. *Journal of Comparative and Physiological Psychology*, **98**, 267–75.

Tallamy, D. W. (1982). Age specific maternal defense in *Gargaphia solani* (*Hemiptera*: Tingidae). *Behavioral Ecology and Sociobiology*, **11**, 7–11.

Tallamy, D. W. (1984). Insect parental care. *Bioscience*, **34**, 20–4.

Tallamy, D. W. (1985). 'Egg dumping' in lace bugs (*Gargaphia solani*, Hemiptera: Tingidae). *Behavioral Ecology and Sociobiology*, **17**, 357–62.

Tallamy, D. W. (1986). Age specificity of 'egg dumping' in *Gargaphia solani* (Hemiptera: Tingidae). *Animal Behaviour*, **34**, 599–603.

Tallamy, D. W. & Denno, R. F. (1981). Maternal care in *Gargaphia solani* (Hemiptera: Tingidae). *Animal Behaviour*, **29**, 771–8.

Taylor, G. T. (1979). Reinforcement and intraspecific aggressive behavior. *Behavioral and Neural Biology*, **27**, 1–24.

Taylor, G. T. (1980). Fighting in juvenile rats and the ontogeny of agonistic behavior. *Journal of Comparative and Physiological Psychology*, **94**, 953–61.

Tellegen, A. & Horn, J. M. (1972). Primary aggressive motivation in three inbred strains of mice. *Journal of Comparative and Physiological Psychology*, **78**, 297–304.

Tellegen, A., Horn, J. M. & Legrand, R. G. (1969). Opportunity for aggression as a reinforcer in mice. *Psychonomic Science*, **14**, 104–5.

Terman, C. R. (1984). Presence of the female and aggressive behaviour in male Prairie Deermice (*Peromyscus maniculatus bairdi*). *Animal Behaviour*, **32**, 931–2.

Thompson, T. (1969). Aggressive behaviour of Siamese fighting fish. In *Aggressive Behaviour*, ed. S. Garattini & E. B. Sigg, pp. 15–31. Amsterdam: Excerpta Medica Foundation.

Thompson, W. R. & Wright, J. S. (1979). 'Persistence' in rats: effect of testosterone. *Physiological Psychology*, **7**, 291–4.

Thor, D. H. (1980). Testosterone and persistence of social investigation in laboratory rats. *Journal of Comparative and Physiological Psychology*, **94**, 970–6.

Thouless, C. R. & Guinness, F. E. (1986). Conflict between red deer hinds: the winner always wins. *Animal Behaviour*, **34**, 1166–71.

Thresher, R. (1985). Brood-directed parental aggression and early brood loss in the coral reef fish, *Acanthochromis polyacanthus* (Pomacentridae). *Animal Behaviour*, **33**, 897–907.

Thurmond, J. B. & Lasley, S. M. (1979). The development of territorial-induced intermale agonistic behavior in albino laboratory mice. *Aggressive Behavior*, **5**, 163–71.

Tinbergen, N. (1951). *The Study of Instinct*. London: Oxford University Press.

Tinbergen, N. (1957). The functions of territory. *Bird Study*, **4**, 14–27.

Tinbergen, N. (1963). On the aims and methods of ethology. *Zeitschrift für Tierpsychologie*, **20**, 410–33.

Tinbergen, N. (1965). *The Social Behaviour of Animals*. London: Chapman & Hall.

Tinbergen, N., Broekhuysen, C. J. Feekes, F., Houghton, J. C. W., Kruuk, H. & Szulc, E. (1962). Egg shell removal by the black-headed gull, *Larus ridibundus* L.; a behaviour component of camouflage. *Behaviour*, **19**, 74–117.

Toates, F. M. (1980). *Animal Behaviour: A Systems Approach*. Chichester & New York: Wiley.

Toates, F. M. & Archer, J. (1978). A comparative review of motivational systems using classical control theory. *Animal Behaviour*, **26**, 368–80.

Tokarz, R. R. (1985). Body size as a factor determining dominance in staged encounters between male brown anoles (*Anolis sagrei*). *Animal Behaviour*, **33**, 746–53.

Toro, M. & Silio, L. (1986). Assortment of encounters in the two-strategy game. *Journal of Theoretical Biology*, **118**, 193–204.

Townsend, D. S., Stewart, N. M. & Pough, F. H. (1984). Male parental care and its adaptive significance in a neotropical frog. *Animal Behaviour*, **32**, 421–31.

Townshend, T. J. & Wootton, R. J. (1985). Variation in the mating system of a biparental cichlid fish, *Cichlasoma panamense. Behaviour*, **95**, 181–97.

Trivers, R. L. (1971). The evolution of reciprocal altruism. *Quarterly Review of Biology*, **46**, 35–57.

Trivers, R. L. (1972). Parental investment and sexual selection. In *Sexual Selection and the Descent of Man*, ed. B. Campbell, pp. 136–79. Chicago: Aldine.

Trivers, R. L. (1974). Parent–offspring conflict. *American Zoologist*, **14**, 249–64.

Trivers, R. L. (1985). *Social Evolution*. Menlo Park, California: Benjamin/Cummings.

Turner, G. F. & Huntingford, F. A. (1986). A problem for game theory analysis: assessment and intention in male mouthbreeder contests. *Animal Behaviour*, **34**, 961–70.

Uhrich, J. (1938). The social hierarchy in albino mice. *Journal of Comparative Psychology*, **25**, 373–413.

Ulrich, R. E. (1966). Pain as a cause of aggression. *American Zoologist*, **6**, 643–6.

Ulrich, R. E. & Azrin, N. H. (1962). Reflexive fighting in response to aversive stimulation. *Journal of the Experimental Analysis of Behavior*, **5**, 511–20.

Ulrich, R. E., Dulaney, S., Arnett, M. & Mueller, K. (1973). An experimental analysis of nonhuman and human aggression. In *The Control of Aggression*, ed. J. F. Knutson, pp. 79–111. Chicago: Aldine.

Ulrich, R. E., Johnston, M., Richardson, J. & Wolff, P. (1963). The operant conditioning of fighting behavior in rats. *Psychological Record*, **13**, 465–70.

Valzelli, L. (1969). Aggressive behavior induced by isolation. In *Aggressive Behavior*, ed. S. Garattini & E. Sigg, pp. 70–6. New York: Wiley.

Van de Poll, N. E., Smeets, J., van Oyen, H. G. & Vander Zwan, S. M. (1982). Behavioral consequences of agonistic experience in rats: sex differences and the effects of testosterone. *Journal of Comparative and Physiological Psychology*, **96**, 893–903.

Van Hooff, J. A. R. A. M. (1972). A comparative approach to the phylogyny of laughter and smiling. In *Non-Verbal Communication*, ed. R. A. Hinde, pp. 209–41. Cambridge: Cambridge University Press.

Van Lawick Goodall, J. (1968). The behaviour of free-living champanzees on the Gombe Stream Reserve. *Animal Behaviour Monographs*, **1**, 161–311.

Van Oyen, H. G. (1981). 'Reactions to novel and aversive stimuli in female and male rats.' Doctoral dissertation, University of Utrecht.

Van Rhijn, J. G. (1974). Behavioural dimorphism in male ruffs *Philomachus pugnax* (L.). *Behaviour*, **47**, 153–229.

Van Rhijn, J. G. & Vodegal, R. (1980). Being honest about one's intentions: an ESS for animal conflicts. *Journal of Theoretical Biology*, **85**, 623–41.

Vandenberg, L. G. (1971). The effects of gonadal hormones on the aggressive behaviour of adult golden hamsters. *Animal Behaviour*, **19**, 589–94.

Vehrencamp, S. L. & Bradbury, J. W. (1984). Mating systems and ecology. In *Behavioural Ecology* (2nd edn), ed. J. R. Krebs & N. B. Davies, pp. 251–78. Oxford: Blackwell Scientific.

Vernon, W. & Ulrich, R. E. (1966). Classical conditioning of pain-elicited aggression. *Science*, **152**, 668–9.

Verrell, P. A. (1984). Sexual interference and sexual defense in the smooth newt, *Triturus vulgaris* (Amphibia, Urodela, Salamandridae). *Zeitschrift für Tierpsychologie*, **66**, 242–54.

Villars, T. A. (1983). Hormones and aggressive behavior in teleost fish. In *Hormones and Aggressive Behavior*, ed. B. B. Svare, pp. 407–33. New York: Plenum.

Vom Saal, F. S. (1984). The intrauterine position phenomenon: effects on physiology, aggressive behavior and population dynamics in house mice. In *Biological Perspectives on Aggression*, ed. K. J. Flannelly, R. J. Blanchard & D. C. Blanchard, pp. 135–79. New York: Liss.

von Neumann, J. & Morganstern, O. (1944). *Theory of Games and Economic Behavior*. Princeton: Princeton University Press.

Vowles, D. M. & Harwood, D. (1966). The effect of exogenous hormones on aggressive behaviour in the ring dove (*Streotopelia risoria*). *Journal of Endocrinology*, **36**, 35–51.

Waage, J. K. (1979). Dual function of the damselfly penis: sperm removal and transfer. *Science*, **203** 916–18.

Waddington, C. H. (1957). *The Strategy of Genes*. London: Allen & Unwin.

Waddintgon, C. H. (1977). *Tools for Thought*. London: Palladin/Jonathan Cape.

Wagner, C. K., Kingsley, C. & Svare, B. B. (1986). Mice: postpartum aggression is elevated following prenatal progesterone exposure. *Hormones and Behavior*, **20**, 212–21.

Walther, F. R. (1977). Sex and activity dependency of distances between

Thompson's gazelles (*Gazella thompsoni* Gunther). *Animal Behaviour*, **25**, 713–19.

Wells, K. D. (1977). The social behaviour of anuran amphibians. *Animal Behaviour*, **25**, 666–93.

Wells, K. D. (1981). Parental behavior of male and female frogs. In *Natural Selection and Social Behavior: Recent Research and New Evidence*, ed. R. D. Alexander & D. W. Tinkle, pp. 184–97. New York: Chiron Press.

Werren, J. H., Gross, M. R. & Shine, R. (1980). Paternity and the evolution of male parental care. *Journal of Theoretical Biology*, **82**, 619–31.

White, M., Mayo, S. & Edwards, D. A. (1969). Fighting in female mice as a function of the size of the opponent. *Psychonomic Science*, **16**, 14–15.

Wiepkema, P. R. (1977). Agressief gedrag als regalsystem. In *Agressief Gedrag: Oorzaken en Functies*, ed. P. R. Wiepkema & J. A. R. A. M. van Hooff, pp. 69–78. Utrecht: Bohn, Scheltema & Holkema.

Wilcox, R. S. & Ruckdeschel, T. (1982). Food threshold territoriality in a water strider (*Gerris remigis*). *Behavioral Ecology and Sociobiology*, **11**, 85–90.

Williams, G. C. (1966). *Adaptation and Natural Selection*. Princeton, New Jersey: Princeton University Press.

Williams, G. C. (1975). *Sex and Evolution*. Princeton, New Jersey: Princeton University Press.

Wilson, E. O. (1971). *The Insect Societies*. Cambridge, Massachusetts: Belknap (Harvard University Press).

Wilson, E. O. (1975). *Sociobiology: The New Synthesis*. Cambridge, Massachusetts: Harvard University Press.

Wilson, P. & Franklin, W. L. (1985). Male group dynamics and intermale aggression of Guanacos in Southern Chile. *Zeitschrift für Tierpsychologie*, **69**, 305–28.

Wingfield, J. C. (1985). Short-term changes in plasma levels of hormones during establishment and defense of a breeding territory in male song sparrows, *Melospiza melodia*. *Hormones and Behavior*, **19**, 174–87.

Wolff, J. O. (1985). Maternal aggression as a deterrent to infanticide in *Peromyscus leucopus* and *P. maniculatus*. *Animal Behaviour*, **33**, 117–23.

Wootton, R. J. (1970). Aggression in the early phases of the reproductive cycle of the male three-spined stickleback (*Gasterosteus aculeatus*). *Animal Behaviour*, **18**, 740–6.

Wrangham, R. W. (1980). An ecological model of female-bonded primate groups. *Behaviour*, **75**, 262–300.

Wuensch, K. L. & Cooper, A. J. (1981). Preweaning paternal presence and later aggressiveness in male *Mus musculus*. *Behavioral Biology*, **32**, 510–15.

Wynne Edwards, V. C. (1962). *Animal Dispersion in Relation to Social Behaviour*. Edinburgh: Oliver & Boyd.

Yasukawa, K. (1979). A fair advantage in animal confrontations. *New Scientist*, **84**, 366–8.

Yasukawa, K. & Searcy, W. A. (1982). Aggression in female red-winged blackbirds: a strategy to ensure male parental investment. *Behavioral Ecology and Sociobiology*, **11**, 13–17.

Yen, H. C. Y., Day, C. A. & Sigg, E. B. (1962). Influence of endocrine factors on development of fighting behavior in rodents. *Pharmacologist*, **4**, 173.

Zack, S. (1975). A description and analysis of agonistic behavior patterns in an opisthobranch mollusc, *Hermissenda crassicornis*. *Behaviour*, **53**, 238–67.

Zahavi, A. (1971). The social behaviour of the white wagtail *Motacilla alba alba* wintering in Israel. *The Ibis*, **113**, 203–11.

Zahler, R. S. & Sussmann, H. J. (1977). Claims and accomplishments of applied catastrophe theory. *Nature*, **269**, 759–63.

Zeeman, E. C. (1976). Catastrophe theory. *Scientific American*, **234**, 65–83.

AUTHOR INDEX

SUBJECT INDEX